Elementary Physics

ENERGY

Its Forms, Changes, & Functions

Tom DeRosa
Carolyn Reeves

ENERGY
Its Forms, Changes, & Functions

Tom DeRosa
Carolyn Reeves

First Printing: August 2009

Master Books®
P.O. Box 726
Green Forest, AR 72638

Printed in the United States of America

Cover Design by Diana Bogardus
Interior Design by Terry White

ISBN 10: 0-89051-570-0
ISBN 13: 978-0-89051-570-9

Library of Congress number: 2009929835

All Scripture references are New International Version unless otherwise noted.

Please visit our website for other great titles: www.masterbooks.net

Investigate the Possibilities

TABLE OF CONTENTS

About the Authors

Tom DeRosa, as an experienced science educator and a committed creationist, has incorporated both his passions in the founding and the directing of the Creation Studies Institute, a growing national creation organization whose chief focus is education. His wealth of experience in the public school, Christian school, and homeschool markets for over 35 years has given special insights into what really works in engaging young minds. He holds a master's degree in education, with the emphasis of science curriculum. He is an author and sought-out, enthusiastic creation speaker who has a genuine love for the education of our next generation.

Carolyn Reeves is especially skilled at creating ways to help students develop a greater understanding of not just scientific concepts, but also how these are applied within the world around us. Carolyn retired after a 30-year career as a science teacher, finished a doctoral degree in science education, and began a new venture as a writer and an educational consultant. She and her husband make their home in Oxford, Mississippi, where they are active members of North Oxford Baptist Church. The Reeves have three children, three in-law children, and ten grandchildren.

# Photo Credits
L = left, TL = top left, BL = bottom left, R = right, TR = top right, BR = bottom right, TC = top center, BC =bottom center

All photos are from Shutterstock unless otherwise noted.

Clipart.com: pg. 34-BR, 35.

Cornelia Bularca: pg. 7-TR, 11-TR, 15-TR, 19-TR, 23-TR, 27-TR, 31-TR, 37-TR, 41-TR, 45-TR, 49-TR, 53-TR, 57-TR, 61-TR, 69-TR, 73-TR, p79-TR, 83-TR.

Created By Design pg. 17-BR, 22-C, 44-C, 50-TR.

Istock.com: pg. 4-BL, 10-BR, 16-BR, 28-TC, 42-L.

NASA: pg. 21-BR, 28-B.

Wikimedia: pg. 6-L, 21, 24, 28-R, 42, 60-L, 68-L, 86-L, 87-TR.

INTRODUCTION

The world has tremendous energy needs, both essential needs and those for desired lifestyles. Energy makes our cars go and helps us stay warm in the winter and cool in the summer. It gives us cold drinks or hot french fries, lights to see, and tunes to hear.

One of the amazing things about energy is that you can never destroy it and you can never create it. You might think the light in a room disappears at the flick of an electric switch, but the light energy is just changed into heat and mechanical energy.

Scientists define energy as the ability to do work. However, to be useful, energy must be harnessed and directed. For example, electrical energy has little practical value when it is in the form of electricity. There is a huge demand for electricity because it is easy to change electrical energy into other forms of energy, like heat or light or sound. Engineers have developed numerous ways to change energy into forms we can use to make life easier and more entertaining!

We will be looking at some of the basic forms of energy — light, heat, chemical, electrical, mechanical, and nuclear. We will also be looking at natural fuels and household electricity.

There are two main categories of fuels: nonrenewable fuels that can't be recreated in a small amount of time and renewable fuels that can be produced in a short period of time as long as we have sun, wind, water, plant life, and heat inside the earth.

People are using massive amounts of the earth's nonrenewable sources of oil, natural gas, and coal, as well as uranium. Our need for energy sometimes conflicts with our responsibility to take care of our environment or to prevent air, water, and soil pollution. There is also a danger that we will squander our energy resources instead of conserving them.

The need for solutions to our energy needs is huge. Maybe some of you will be part of the solution. You may run a wind farm, produce new kinds of fuels from plants or garbage, or discover some brand-new source of energy.

You may also want to become a scientist and help unravel the mysteries of the origins of the nonrenewable energy sources. Some scientists are challenging the popular theory that these energy sources were created millions and millions of years ago by gradual processes. They believe instead that major catastrophes were the primary way in which huge amounts of vegetation and other living things became the world's supply of oil, natural gas, and coal. Could that have something to do with Noah's Flood?

Scientists

Aristotle .... (384 B.C.–322 B.C.)

Galileo .... (1564–1642)

Michael Faraday .... (1791–1867)

Joseph Henry .... (1797–1878)

Samuel Morse .... (1791–1872)

Hans Christian Oersted .... (1777–1851)

Lord Kelvin .... (1824–1907)

Enrica Fermi .... (1901–1954)

Albert Einstein .... (1879–1955)

Lise Meitner .... (1878–1968)

Niels Bohr .... (1885–1962)

HOW TO USE THIS BOOK

Each investigation gives students a chance to learn more about some part of God's creation. To get the most out of this book, students should do each section in order. Many science educators believe science is best learned when students begin with an investigation that raises questions about why or how things happen, rather than beginning with the explanation. The learning progression recommended for this book is: engage, investigate, explain, apply, expand, and assess. In each lesson, students will be introduced to something that is interesting, they will do an investigation, they will find a scientific explanation for what happened, they will be able to apply this knowledge to other situations and ideas, they will have opportunities to expand what they learned, and there will be multiple assessments.

Think about This (Engage) — Students should make a note of what they know or have experienced about the topic. If this is a new topic, they could write some questions about what they would like to learn.

The Investigative Problem(s) — Students should be sure to read this so they will know what to be looking for during the investigation.

Gather These Things — Having everything ready before starting the investigation will help students be more organized and ready to begin.

Procedures and Observations (Investigate) — Students should first follow the instructions given and make observations of what happens. There will usually be opportunities for students to be more creative later.

The Science Stuff (Explain) — This section will help students understand the science behind what they observed in the investigation. The explanations will make more sense if they do the investigation first.

Making Connections (Apply) — Knowledge becomes more permanent and meaningful when it is related to other situations and ideas.

Dig Deeper (Expand) — This is an opportunity for students to expand what they have learned. Since different students will have different interests, having choices in topics and learning styles is very motivating. All students should aim to complete one "Dig Deeper" project each week, but the teacher may want older students to do more. Generally, students will do at least one project from each lesson, but this is not essential. It is all right for students to do more than one project from one lesson and none from another.

What Did You Learn? (Assessment) — The questions, the investigations, and the projects are all different types of assessments. For "What Did You Learn?" questions, students should first look for answers on their own, but they should be sure to correct answers that might not be accurate.

Additional opportunities for creative projects and contests are found throughout the book. For grading purposes, they can be counted as extra credit or like a "Dig Deeper" project.

Nurture Wisdom and Expression

Each book contains information about early scientists and engineers. Students need to see that they were regular people who had personal dreams and who struggled with problems that came into their lives. Students may be surprised to realize how many of the early scientists believed that understanding the natural world gave glory to God and showed His wisdom and power.

In addition to the science part, students will find creation apologetics and Bible mini-lessons. The apologetics will clear up many of the misconceptions students have about what science is and how it works. Both the apologetics and Bible lessons should lead to worthwhile discussions that will help students as they form their personal worldviews.

Students with artistic and other creative interests will have opportunities to express themselves. For example, some of the apologetics are written in narrative form and are suitable for drama presentations. As scientists are introduced and researched, students can also present what they have learned as time-dated interviews or news accounts. Remember, if the scientists are included in a drama presentation, they should be represented as professionals, not as stereotyped, weird-looking people.

Where Exactly Does Energy Go?

Think about This

Ella understands that light is a form of energy, but she is having trouble with the idea that light energy cannot be created or destroyed.

German-born Albert Einstein was awarded the 1921 Nobel Prize in physics. His studies of light transformation helped to base his discovery of the photoelectric effect.

"Look," she told her aunt, who is a science teacher. "When I flip the switch and turn off the lights, I cause all the lights in the room to go away." She demonstrated and made the room very dark.

"Now look what happens when I turn the light switch back on. The room fills with light again. Didn't I just create and destroy the light in the room?" she asked.

"No, you certainly did not," her aunt said. "All you did was demonstrate how energy can change from one form into another."

Let's look at some examples of how energy changes from one form to another in this lesson.

The Investigative Problems

What are examples of energy?

Can one form of energy change into another form of energy?

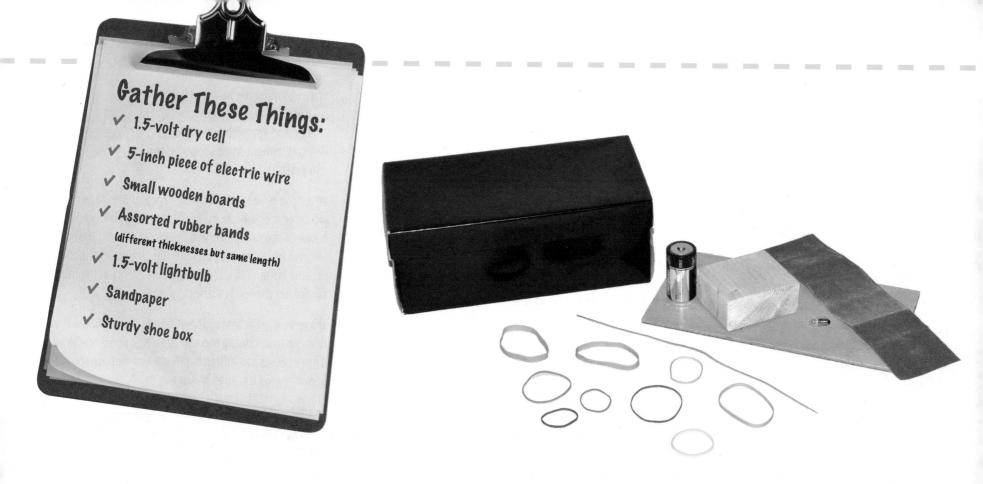

Gather These Things:

✔ 1.5-volt dry cell

✔ 5-inch piece of electric wire

✔ Small wooden boards

✔ Assorted rubber bands
(different thicknesses but same length)

✔ 1.5-volt lightbulb

✔ Sandpaper

✔ Sturdy shoe box

Procedure & Observations

1. Electric energy to light and heat energy: Take a 1.5-volt dry cell, a five-inch wire, and a light bulb. Test different combinations until you get the light bulb to come on. Show your teacher when you are successful. Make a drawing to show how you connected everything.

   Feel the light bulb. Can you tell if it has gotten any warmer? (Note: This is a small amount of heat and it may not be easy to detect.)

2. Mechanical energy to heat energy: Rub a piece of sandpaper quickly over a board several times. Feel the sandpaper and the board. What kind of energy is produced?

3. Mechanical energy to sound energy: Remove the cover from a sturdy box and cut three grooves on opposite edges of the box. Now choose three rubber bands of equal length, but each with a different thickness. Stretch the rubber bands around the box, fitting each into one of the grooves. Pluck each rubber band. Observe that it is vibrating. Listen for a sound. Repeat for each rubber band. Compare the pitch made by the different rubber bands. Record your observations.

The Science Stuff

Energy is what enables matter to move or to change. Energy is found in many different forms, such as heat, light, electricity, mechanical (the energy in moving things), sound, nuclear, and chemical. One form of energy can be changed into another form of energy. Still, the total amount of energy never changes. This means that energy cannot be created or destroyed. These ideas are expressed in one of the most important laws in all of science — the law of conservation of energy.

These activities illustrate some of the main forms of energy. Each activity shows one form of energy being changed into another form of energy. Electrical energy changed into light and heat, mechanical energy changed into heat, and mechanical energy changed into sound.

In the first activity, when the equipment was wired together correctly, an electric circuit was completed. An electric current then moved through the dry cell, wires, and light bulb. As the electric current moved through the light bulb, electric energy changed into light energy and heat energy.

This activity illustrates another important concept about energy. It shows that energy can be transferred from one place to another. Much of the earth's energy is transferred from the sun to the earth.

Remember the conversation between Ella and her aunt? When Ella flipped the light switch, the electric current began to move through the wires and the light bulb. Inside the light bulb, electric energy changed into light and heat energy, which is the same thing that happened in your activity with

electricity. When she turned the lights off, the objects in the room absorbed the heat and light energy. (This is a small amount of energy, and you probably couldn't detect it without some sophisticated equipment.)

When you rubbed a board with sandpaper, your motion produced mechanical energy. This motion produced friction between the sandpaper and the wood, causing the molecules to move faster. As a result, both the sandpaper and the wood became hotter. Thus, the mechanical energy of the moving sandpaper changed into heat energy.

You were also the source of motion when you plucked the tight rubber bands, causing them to vibrate. Sound is produced when a force causes something to vibrate and produce sound waves. Sound energy is carried in waves.

Making Connections

Another way in which mechanical energy can produce sound waves is by tapping on a table. Tapping on the table causes the table to vibrate in the same way plucking on the rubber bands caused them to vibrate. Sound waves actually travel faster through the table than through the air. You can put your ear next to the table and hear the tapping sounds clearly. You can also raise your head and hear the sounds as the sound waves pass through the table and then through the air.

When electrical energy passes through a light bulb, it is changed into light energy and heat energy. Even though the heat energy is unwanted, it is still part of the electric bill. Engineers try to design light bulbs that increase the amount of light and decrease the amount of heat produced. Some progress has been made, but light bulbs continue to produce unwanted heat.

Dig Deeper

Start with the energy being given off from a TV or a radio in your home. Try to figure out where this energy comes from. See how far back you can trace the energy changes. This gets a little complicated, so get a good reference book to help you.

What is the difference between an electric motor and an electric generator? They basically contain the same parts and are built the same way. However, an electric motor changes electric energy into mechanical energy, and an electric generator changes mechanical energy into electric energy.

In 1905, Albert Einstein proposed a theory that altered the law of conservation of energy. He said that matter can be changed into energy, and energy can be changed into matter, but the total amount of matter and energy in the universe remains the same. How was Einstein's theory shown to be true?

What Did You Learn?

1. Give two examples of how one form of energy can change into heat energy. Give another example of an energy change.

2. List two ways in which energy does work for us.

3. The following list contains examples of forces, properties of matter, and forms of energy. Underline all the examples of forms of energy: inertia, light, heat, density, buoyancy, electricity, lift, weight, chemical, push, and nuclear.

4. Define mechanical energy and give an example.

5. What kind of energy can be quickly provided by a battery?

6. What is the law of conservation of energy?

7. Give an example of when an unwanted form of energy is produced in a device.

8. What happens to a roomful of light on a dark night when the lights are turned off?

9. Was energy transferred from the battery to the light bulb when an electric circuit was completed?

Stored or Active?

Think about This   Not all forms of energy are in an active state. Sometimes energy is stored or in a position to become active at a later time. Christopher loves to ride a sled down a snow-covered incline. His friend John Scott loves to play on the water slides. Actually, both activities illustrate how energy can be stored as potential energy and converted into active kinetic energy.

The Investigative Problems
Does energy exist in potential and kinetic forms?
Can these two forms of energy be changed from one form into another?

Gather These Things:

✓ A few feet of pipe foam insulation (3/4-in. to 7/8-in. inside diameter, no thicker than 3/8 in.)

✓ Various sizes of objects to make hills (such as cans)

✓ Empty paper towel rolls

✓ Tape

✓ Marbles

✓ Meter stick or measuring tape

✓ Scissors to cut insulation

✓ Other creative supplies

Procedure & Observations

Cut open the foam insulation to make two pieces of track. Tape one end of the track to a chair or a stack of books and put a small cylindrical can under the track near the middle to make a hill. Release a marble from the top of the track and see if it has enough energy to make it over the hill.

Make the hill a little taller than the beginning of the track and see if the marble has enough energy to make it over the hill. Increase the height of the starting place until the marble has enough energy to roll over the hill. Make drawings of two designs that work and one design that doesn't work. Measure and include the heights of your hills and the starting positions in your drawings.

Now design a long roller coaster track for a marble to follow. Use both pieces of insulation and tape as needed. Include hills, valleys, and curves. You may include loops and barrel curves if you wish. Construct a track according to your design. Use tape if needed to make it more secure. Test your track by releasing a marble at the starting point. Did it make it all the way to the end of your track?

Make a list of the curves, hills, and other features where the marble stopped or jumped the track. Try to think of a reason why it did this in each instance.

If the marble didn't make it to the end of the track or you simply want to make it more exciting, adjust or redesign your track. You may find that the track can be pushed through empty toilet paper rolls to make a tunnel. The paper rolls are also easier to tape down than the track. The directions are not specific. This is to encourage you to be creative and to try some things that might increase the efficiency of your track. Feel free to put some creative artwork with this investigation.

Test your design again by releasing the marble at the starting point. Make a drawing or diagram of your final track.

The Science Stuff

Energy can be stored in such a way that it can be used at a later time. This form of energy is known as potential energy. Energy can also be in a form that is being actively used. This is known as kinetic energy.

Specific kinds of energy, such as heat, light, electricity, chemical, sound, mechanical, or nuclear, can exist as both potential and kinetic forms of energy.

When the marble is sitting motionless at the starting position, all its energy is potential. When the marble begins to roll down the hill and gets closer to ground level, it has less and less potential energy. The marble has the most potential energy at the highest point on the track. It has the least potential energy at the lowest level.

At the same time the marble is speeding down the hill, it is gaining kinetic energy. The marble has the most kinetic energy at the point where its speed is greatest and the least where its speed is zero. The kinetic energy becomes higher going down the hills and gets lower going up a hill.

As the marble makes its way around the track, its potential energy and kinetic energy are constantly changing. Sometimes potential energy is changed into kinetic energy. Sometimes kinetic energy is changed into potential energy.

In order for the marble to get to the starting place, your body had to expend energy to lift it and place it there. The higher the marble is at the start, the more potential energy it will have. The more potential energy the marble has at the start, the more energy it will have on its route to the end of the track.

A marble will travel farther and faster each time it starts from a higher position, but it cannot roll up a hill that is higher than its starting position. If your first hill is higher than the starting point, the marble will not have enough energy to make it up the hill.

The force of gravity (or the weight of the marble) causes the marble to be pulled down the ramp. The marble is also affected by friction and other forces that push back on it. Friction may keep the marble from rolling as high as you might have expected it to.

There are many ways an object can acquire potential energy. Water that is behind a dam has a great deal of potential mechanical energy. The higher the dam, the more potential energy the water will have. A stretched rubber band would have potential mechanical energy. It will change into kinetic mechanical energy if it is used to propel a paper wad through the air. A battery or a tank full of gasoline would have potential or stored chemical energy. The energy in the gasoline will eventually be converted into kinetic mechanical energy as it makes the car move.

In the activities that were done in the first lesson, you saw other examples of stored (potential) energy. For example, the disconnected battery you used contained stored potential energy. The stretched rubber band had potential energy before you plucked it.

You also observed examples of energy in use (kinetic energy). For example, the stored energy in the battery was quickly changed into electric kinetic energy when you made a complete circuit. Producing heat by rubbing a board with sandpaper involved kinetic mechanical energy. The rubber band had kinetic mechanical energy when you plucked it and caused it to vibrate.

Making Connections

Energy changes occur all the time. Sometimes they are purposeful and other times they just occur naturally. Christopher and John Scott experience energy changes every time they sled down a snow-covered incline or play on the water slides. They start from a position where there is enough potential gravitational energy to carry them on a fun ride. Christopher found a place to start where he can sled down one hill and almost all the way up another hill. By walking up to the top of the second hill, he can sled back to his starting place.

Archers who shoot arrows with a bow notice that the farther the bowstring is pulled back, the farther the arrow will travel. This is because pulling the string farther back gives the string more potential energy.

It is important to be able to store energy so it can be used later. Cars, for example, contain a battery for storing energy that can provide the electricity needed to start the engine. When you have a tank full of gasoline in your car, you have a tank full of stored chemical energy, ready to be used.

Nuclear energy is another form of energy. Potential chemical and nuclear energy are stored in atoms. Chemical energy is released when electrons (negative particles) change positions. Nuclear energy can be released when there are changes in the nucleus of the atom. Nuclear energy is thousands of times more powerful than chemical energy.

Dig Deeper

Devise an experiment where you shoot ping-pong balls with a rubber band slingshot. Measure how far a ball will travel according to how far the rubber band is pulled back. Be sure the distance the rubber band is pulled is the only variable in your experiment.

As the price of gasoline has increased in the past few years, there has been renewed interest in electric cars and hybrid cars (cars that can switch back and forth between electric power and gasoline power). One of the problems with electric cars is that the batteries don't last long before they have to be recharged. Do some research on batteries for electric cars.

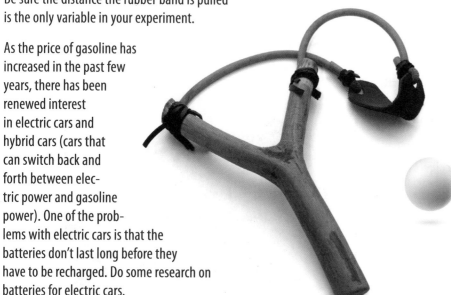

What Did You Learn?

1. Give two examples of energy that is stored in such a way that it can be used at a later time.

2. Does a tank full of gasoline contain potential or kinetic energy?

3. At what point on a roller coaster ride is the potential energy the greatest?

4. At what point on a roller coaster ride is the kinetic energy the greatest?

5. How can an archer increase the potential energy of an arrow that is about to be shot?

6. Which form of energy is more powerful — nuclear or chemical?

7. How can a marble be given more potential energy at the beginning of the track?

8. What force prevented the marble from rolling a little higher than it did?

Pause and Think:

What happens to our cars and air conditioners if fuel supplies are used up? When a dry twig is lit, the chemical energy that is stored inside the twig changes into heat, light, and even sound energy. The same thing happens when coal, oil, and gas products are burned. These resources are the remains of once-living plants and animals. They still contain stored chemical energy. The gasoline that goes into our cars comes from underground oil.

Most cars and trucks today have internal combustion heat engines. The fuel is ignited by a spark from a spark plug (or by compression, in the case of diesel engines). The chemical energy in the fuel changes into heat, and the gases trapped in the engine begin to expand and push. This means that heat energy changed into mechanical energy. The engine is designed to use this pushing force to make the car go.

Will we ever run out of oil to make gasoline for our cars? What about

resources that provide fuel that make our heaters and air conditioners work? That is a scary thought, since coal, oil, and gas are considered nonrenewable fuels.

The United States still has huge supplies of these fuels, but eventually we will run out of them if we continue the present rate of usage. Scientists and engineers are searching for other sources of energy and fuels. A few of the more promising solutions include turning used cooking oil into diesel fuel, making oil from garbage and other biomass, making long-lasting batteries to power cars, and finding ways to better conserve the resources we have.

To Dig Deeper about this important problem, see if you can find more information about what is being done to conserve our natural coal, oil, and gas resources. Try to find alternative ways to provide energy that will make our cars, heaters, and air conditioners work. For now, just summarize several ideas. You will have opportunities to research specific methods and ideas later.

COOKING OIL

13

Light: Reflected and Absorbed

Think about This
Beams of light are all around us. If you could see what they were doing, you would see some light beams hitting objects and bouncing off; others, passing through objects; and others, changing into heat as they hit objects. If they came from the sun, they had to travel through several million miles of space before bumping into the earth and the air around it.

In this lesson, we will investigate two things that can happen to light when it hits an object — **reflection** and **absorption**.

The Investigative Problems
What happens to light when it hits an object if it doesn't go through the object?

Gather These Things:

- ✔ Two small, flat mirrors
- ✔ One large, highly polished tablespoon
- ✔ Clothespins or other clamps
- ✔ Magnifying lens (convex lens)

Procedure & Observations

Work with a partner. Stand a flat mirror in an upright position on a flat table. Use clothespins or other clamps to make it stand upright without holding it. Position yourself and the mirror so that you are looking in the mirror from one side. You will not be able to see yourself in the mirror from this position. Have your partner move around until you can see him/her. What does your partner see at the same time you see your partner?

Change your position so that you can still see the mirror, but not yourself. Estimate where your partner will have to be in order for you to see him/her in the mirror. Make a diagram of you, the mirror, and your partner. Show the path of a beam of light coming from your partner to the mirror to your eyes.

Sit with your back to a window and hold a flat mirror in front of you. Tilt the mirror away from you until you can see the ceiling. Take a second mirror and hold it above the first one at an angle that lets you see outside the window. Move the mirrors slowly until you get them adjusted. Describe what you see.

Find a highly polished tablespoon. Look at yourself using the inside of the spoon. Move the spoon back and forth. What kind of images do you see at different distances? Look at yourself using the back of the spoon. How do you look?

Take a piece of white paper and make a 1-cm. black circle with your pencil on the paper. Go outside and use a magnifying lens (a convex lens) to focus sunlight on the white paper. Note what happens. Now use the lens to focus sunlight on the black circle. Note what happens. Do not look directly at the bright light produced by the lens.

window

beam of light

mirror

mirror

Light travels in a straight line and is reflected each time it hits a mirror. This is how a periscope works.

15

The Science Stuff

A beam of light will travel in a straight line unless it hits another object or enters a different object. When light travels through the air and hits a flat, shiny mirror, it will be reflected in a very predictable way. A beam of light that hits a flat mirror at an angle will be reflected at the same angle. If it hits the mirror at a 30 degree angle, it will leave the mirror at a 30-degree angle.

The behavior of light is much like a smooth ball rolling into a flat, smooth surface. Unless the ball is spinning, the ball will leave at the same angle at which it came.

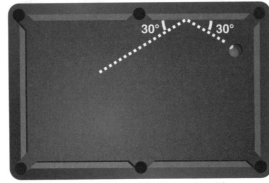

The path of the pool ball is predictable.

If a beam of light hits a rough surface, it will not be reflected in this way. Light will scatter in many different directions after hitting a rough surface.

Light can also be reflected inward or outward from curved, shiny surfaces. Mirrors that curve away from you (concave shape) bring the light to a focus, similar to what you saw with a magnifying (convex) lens. Your image could be small or large, right side up or upside-down, depending on how far away the spoon is. Mirrors that curve toward you (convex shape) produce a small, upright image.

When light hits an object, it is sometimes reflected from the object. Second, light may be absorbed and transformed into heat energy. Third, light may pass through some substances. We will investigate this third possibility in the next lesson.

When light is focused on a white piece of paper with a magnifying glass, you will probably not see a change in the paper. But when the light is focused on a black piece of paper, the paper will probably begin to smoke or flame up. This illustrates the difference in reflection of light and absorption of light. The white paper reflected the light. The black paper absorbed the light and changed it into heat energy.

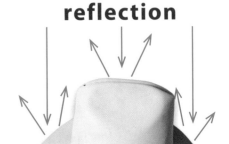

absorption

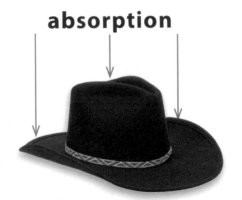

reflection

Making Connections

Black-colored clothing is warmer than similar white-colored clothing. The black clothes do not reflect light. They absorb light, much of which will eventually change into heat. The white clothes reflect light. This is the reason darker colors are a good choice for winter and lighter colors are a good choice for summer. You may have also noticed that walking barefoot on a white beach can be pleasant in the summer, but stepping onto a nearby dark-colored pavement can be painfully hot.

Concave reflectors are used in headlights of cars to reflect a strong beam of light straight ahead and keep it from scattering. Convex mirrors are used in side-installed rear view mirrors to give the driver a wider view of what's behind the car. There may be a warning that what the driver sees in the mirror is closer than it appears. This is because the image seen in this mirror is actually smaller than what would be seen in a flat mirror.

Dig Deeper

Find directions for how to make a periscope. Then assemble the materials and build a periscope that will let you see around a corner.

Kaleidoscopes are fascinating devices. The beautiful images you see in them are the result of the way three mirrors reflect a variety of colored objects inside. With a little help in getting the right kinds of mirrors, you can make your own. Do a search on the Internet for how to make a kaleidoscope. Assemble the equipment you need and make your own kaleidoscope. Share it with others.

What Did You Learn?

1. When light is absorbed by an object, what form of energy will much of it eventually change into?

2. If light strikes a flat mirror at a 45-degree angle, at what angle will it be reflected off the mirror?

3. What happens when light hits a white object?

4. What happens when light hits a black object?

5. Explain why white clothes are cooler in the summer than black clothes.

6. Do mirrors that are curved toward you magnify things or make them appear smaller?

7. What are three things that can happen to a beam of light that hits an object?

8. Why are convex mirrors used in side-installed rear view car mirrors?

Light and Lenses

Think about This

Reading stones, spectacles (eyeglasses), and other uses of lenses were used all over the world before anyone realized that some very neat instruments could be built with a tube and a combination of lenses. There is a credible story that a spectacle maker let two of his children play with some of his lenses, and they found a way to combine them and magnify things. Whoever made this discovery probably had no idea how important it was. It led to the invention of telescopes and microscopes, which opened up whole new fields of science. Early scientists were amazed when they looked into the sky with telescopes and saw things like moons around Jupiter and mountains on the moon. Later, biologists were equally amazed as they looked into microscopes and saw tiny living organisms in a drop of water.

The teacher will demonstrate some of the basic principles about how light passes through a convex lens. Dim the lights in the room. Light a candle. Place a magnifying (convex) lens between the candle and a white sheet of paper. Move the lens slowly back and forth until you can see an image of the candle flame. Is it right side up or upside-down? Move the paper closer and try to refocus the flame. Move the paper farther back and try to refocus the flame. You should be able to see different size images of the flame. Are all the images upside down?

The Investigative Problems
What happens to light when it goes through a convex lens?

Gather These Things:

- ✓ Two convex lenses (magnifying glasses)
- ✓ Shallow (opaque) bowl of water
- ✓ Matches
- ✓ Penny
- ✓ Candle
- ✓ Pencil
- ✓ Glass of water

Procedure & Observations

Work with a partner. Place a penny in a shallow cereal bowl, one you can't see through. Your partner should step back until the penny is no longer visible. Now slowly pour water into the bowl while your partner stands still and keeps looking at the bowl.

Does the penny become visible? Change positions with your partner and repeat this activity.

Put a pencil in a glass of water. It should be about halfway in the water and halfway out of the water. Observe the pencil from different angles. Describe how the pencil looks.

Place a magnifying lens (convex lens) near your eye, and look through it at an object that is at least five meters away. Describe how the object looks. Slowly move the lens farther away from your eye as you continue to look at the object through the lens. Does the object look larger or smaller as you move the lens farther away?

Continue to move the lens farther away from the object. Is there a point where you can't see it because it is too blurry?

Continue to move the lens farther away from the picture. Is there a point where the object looks upside-down? Does the size of the object look smaller or larger?

Continue to look through the lens at your object with your arm fully extended. Now take another magnifying (convex) lens and hold it near your eye. Look through both lenses at the same time. Move the arm that is extended very slowly toward your eye. Stop when you see a focused image. Describe the focused image.

Making Connections
Look at the drawings of a simple camera and a human eye. Note the similarities:

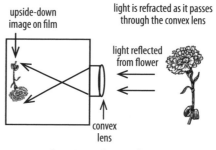

Drawing of a simple camera

Drawing of a human eye

Light enters both the camera and the eye through a small opening. They both contain a convex lens that focuses the light into an image that is smaller and upside-down. If the image is blurry, the picture made by a camera will also be blurry. If the image that forms on our retina is blurry, the things we see may be out of focus, too. The brain interprets the upside-down image on the retina so that we see things right side up!

The Science Stuff

Light travels in a straight line unless its path is interrupted by something. When light hits an object, it can be reflected off the object; it can be absorbed and changed into heat energy; or it can go through the object.

When light goes through water or a lens, it changes directions at the beginning of a new material. The pencil looks bent, because light leaves the part of the pencil in the air and travels straight to your eye. Light from the part of the pencil in the water travels straight until it hits the air. But it makes a sudden change in its direction at the point where the air and the water meet.

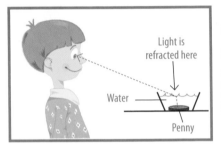

Light is refracted here

Water

Penny

The penny became visible when water was added, because beams of light from the penny changed direction as they left the water and hit the air. They then traveled in a straight line through the air to your eyes.

A beam of light changes speed when it enters a new material. Light behaves somewhat like what would happen to a car that is traveling on a paved highway. If the two right tires leave the pavement and begin traveling through mud, the car would be pulled to the right. This is because the left tires would be gripping the road and, in effect, turning faster than the right tires.

As you looked at an object using a lens that was far away from your eye, you probably saw an upside-down image that was blurry. The blurry image became sharp and clear when you looked through a second lens that was near your eye. In most telescopes, you will find two convex lens that produce a clear, sharp image that is upside-down. The image can become larger and clearer by using different combinations of lenses.

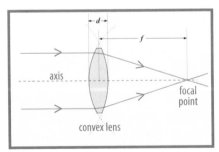

$-d-$

f

axis

focal point

convex lens

The bending of the light by a lens is the key to how eyeglasses, telescopes, microscopes, cameras, movie projectors, and even human eyes work.

Dig Deeper

Early astronomers combined two or more lenses and made telescopes that could be used to study the moon, the planets, and other objects in space. Use reference books or the Internet to find out how to make a simple telescope. Try to construct a telescope that will allow you to see the moon. Look at other objects and describe the images you are able to see. Are the images upside-down? Make a diagram of the telescope you made.

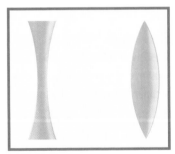

Use reference books or the Internet to see how microscopes are made. Find out how the amount of magnification is determined. Do objects viewed through a microscope look upside-down or right side up? Who were some of the early scientists who used microscopes? As a bonus, you might even try to make a simple microscope that can magnify a small object.

Experiment with combinations of a concave lens and a convex lens to see if you can find a way to focus on an object. Make a diagram to show how you focused on an object. Measure or estimate the distances.

The demonstration at the beginning of the lesson showed what happened to the size of the image as the **object distance** (distance from object to lens) and the **image distance** (distance from image to lens) changed. Use a reference book or the Internet to find out how a movie projector works. Explain how a small film frame can be converted into a large image on a screen in a movie theater. Explain why the film frames have to be put into the projector upside-down.

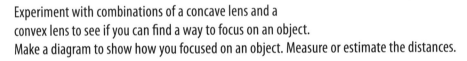

Galileo Takes on Aristotle

Galileo (1564–1642) did not actually invent the first telescope, but when he heard about the discovery he built one for himself. He was able to see such things as the four inner moons of Jupiter and the mountains on the earth's moon with his telescope. This and other evidence convinced him that the Copernican theory of a sun-centered solar system was correct. However, his opinions put him at odds with the ancient teachings of Aristotle, with the Catholic church, and with many other scientists of his day.

Galileo and the Catholic church's famous quarrel over these issues is often misunderstood. It was partly because Galileo believed the sun was the center of the solar system. But the main reason for the quarrel was because Galileo disagreed with Aristotle's methods for studying science and was very critical of the methods he had proposed.

Aristotle, who lived from 384 B.C. to 322 B.C., was a brilliant philosopher. He tried to arrive at answers to important questions by using logic. Aristotle believed the way to learn about something was to first determine its purpose. He believed every object had an innate purpose that it strived to achieve. Over the years, the Catholic church had incorporated many of Aristotle's ideas into what they taught. They especially liked the idea that every object has a purpose.

Galileo insisted that this was no way to study science. He thought observations, experimental methods, and the use of mathematics were the ways to learn about the natural world. He thought that looking for the purpose of an object was a waste of time in scientific study.

Aristotle also taught that the earth was the center of the universe. This was based on ideas that were very logical at the time he lived. The earth did not appear to be moving. It was made up of a dark, heavy mass. The stars were different from the earth because they looked like points of light and did appear to be moving around the earth.

Galileo challenged the methods Aristotle used to study nature. He used the Copernican theory that the earth moved around the sun to make his points. So far, there was still not a lot of scientific evidence to make the average person believe this.

Galileo continued to criticize Aristotle's methods of studying the natural world. Finally, the pope ordered Galileo to return to Rome and answer charges before the Inquisition. Rather than be excommunicated from the church, Galileo recanted (took back) the things he had been saying. He didn't really believe he was wrong, but he stopped speaking publicly about it.

Although Galileo didn't live to see many of the changes that occurred in science, he was responsible for some of the most important changes. The telescope and better scientific methods eventually changed what people thought about the sun and the planets. Galileo's study of forces and motion was applied by Isaac Newton to explain how planets move around the sun. Galileo's ideas helped Newton formulate his laws of motion. Galileo also had an important effect on the methods that would be used by future scientists to study the natural world. The study of nature became much more scientific as a result of what he discovered and taught.

Discuss: Why did the Church try to protect Aristotle's teachings about science?

What Did You Learn?

1. Are the frames for a film that goes into a movie projector upside-down or right side up?

2. Label the parts of the human eye: pupil, convex lens, retina. Where does an image focus inside the eye?

3. When an image focuses inside the eye, is it right side up or upside-down? What part of our body is necessary to interpret this image?

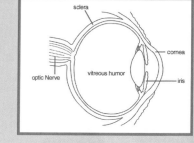

4. What equipment is needed to make a simple telescope?

5. What equipment is needed to make a simple microscope?

6. Compare telescopes and microscopes. Give two ways they are similar.

7. Where is a convex lens the thickest and where is it the thinnest?

8. As light travels through the air, does it go in a straight line?

9. What happens to a beam of light when it leaves the air and enters a glass of water?

Waving the Red, Green, and Blue

Think about This

Take pieces of red, green, and blue cellophane and look at various objects through each. Dim the room and find a poster with red, green, blue, and other colors in it. Shine a bright light on the poster. Observe as your teacher waves red, green, and blue cellophane (taped across three openings of a piece of cardboard) back and forth in front of the light. You may recognize this as a strobe light. What do you think is happening when light passes through the cellophane paper? How will the poster look if the blue cellophane is moved and covers all of the light coming from the flashlight?

The Investigative Problems

What kind of waves are light waves? What happens to light waves that are absorbed? What happens to light waves when we see different colors?

Procedure & Observations

Construct a color spinner. Cut a circle from the cardboard with a 10-cm. diameter. Divide the circle into three sections and color the sections red, green, and blue. Place a two-cm. length of clear tape across the middle of the circle. Punch two holes near the center but off center by about a centimeter, being sure to punch through the tape. Push the string through both holes and tie the string. There should be a loop of string on each side of the spinner. Hold the two loops of the string while someone else turns the spinner several times. When you pull on the string, the circle will spin. Alternately pull and release the string to keep it spinning. What color do you see when the spinner is turning rapidly?

Shine a light through a **prism** or just hold a prism near your eyes and try to find a spectrum of rainbow colors. Name the colors you see in order, starting with red, which has the longest wavelength of visible light.

Work with a partner. Take a metal slinky and stretch it across the table. Put a piece of tape on one of the coils so you can observe the direction of its movement. Your partner should hold the last one or two coils very firmly while you move the slinky and form waves in the slinky. Move your hand quickly to the left and then back to the right. Observe that a **pulse** moves from your hand to your partner. Does it bounce back when it reaches your partner?

Place a piece of tape on one of the slinky coils and observe the direction it moves as you make waves. Now keep moving your hand back and forth in a sideways direction. If you can move your hand at the right speed, smooth, regular **waves** will form. Record your observations or make a drawing of how they look.

The Science Stuff

Spinning the three primary colors of light together produces the same effect as shining the three colors on a screen. The more lights that are added together, the lighter the color. All the primary colors of light mixed together make white light. White is also seen when all the six colors of light seen through a prism are mixed together.

Mixing the primary colors of light together is different from mixing the three primary colors of paints. Paints will get darker as you add more and more different colors.

Green object

When you look through a prism, you can see the different colors of light. The colors you see from ordinary light are red, orange, yellow, green, blue, and violet. Some people use the name ROY G. BIV to help them remember these colors in the correct order. The *i* stands for indigo. At one time indigo was thought to be another color, but it's really the same as violet. Red waves are the longest and violet waves are the shortest of the visible light waves. Each color has a different wavelength. The range of colors you see through a prism is called the visible spectrum.

The colors you see around you depends on which light waves are absorbed and which ones are reflected. For example, when light hits a green object, all the visible waves strike the green object, but only the green waves are reflected and enter your eyes. The other waves are absorbed. The diagram below shows the green waves being reflected and the other waves being absorbed by the object's material. You should be able to make similar diagrams for any color.

You should remember that white light is seen when all the visible light waves are reflected and none are absorbed. Black is seen when all the visible light waves are absorbed and none are reflected. The reason the black area on the paper flamed up more easily than the white areas was because when light hit the black area, it was absorbed and some of it eventually changed into heat. The white area reflected the light.

In the strobe light demonstration, the red cellophane paper absorbed all the visible waves except red; the green absorbed all the visible waves except green; and the blue absorbed all the visible waves except blue. Red objects look red through the red cellophane but will look darker through green or blue cellophane. The same principle applies to light that passes through green and blue cellophane.

Invisible waves that are longer than red are called infrared waves. They are sometimes known as heat waves. Invisible waves that are shorter than violet are called ultraviolet waves. The shorter the waves, the more energy they carry.

The kind of waves you made in the slinky are known as transverse waves. Light waves are transverse waves. Infrared waves, ultraviolet waves, radio waves, and gamma rays are also transverse waves. They differ in wavelength and in the amount of energy they carry. They can travel through space, as well as through air, water, and other kinds of matter.

Transverse waves are made up of crests and troughs. How close together one crest (or one trough) is to another is its wavelength. How tall the crests are or how deep the troughs are determines the intensity of the light.

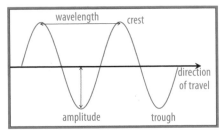

Sometimes light behaves like it is made up of particles instead of waves. Scientists are puzzled by the behavior of light, but they still think that light waves exist.

Making Connections

Color TV sets have three color guns — red, green, and blue which combine colors to produce the colored pictures on your screen. If one of the guns stops working, you only get the colors that two lights can produce.

Dig Deeper

Demonstrate that when red, green, and blue lights are mixed together, white light is produced. Projectors work best, but three good flashlights will also work. Shine one light through the transparent red sheet; one through the blue; and one through the green. Focus the three colors of light on a white surface at the same time, but move them until you get a pattern like this. White (or almost white) should be in the center.

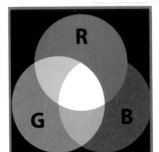

When paints are mixed, the results are different than when lights are mixed. What are the three primary colors of paint? What colors are produced when these colors are combined two at a time?

What Did You Learn?

1. What are the three primary colors of light?

2. Are these the same as the primary colors of paint?

3. What color would you see if equal amounts of the three primary colors of light were shined together on a screen?

4. What is a triangular-shaped piece of glass or plastic that separates the colors of light into the visible spectrum?

5. Of the visible colors of light, which has the longest wavelength? Which has the shortest wavelength?

Red object

6. Explain the diagram to the right, telling which waves are reflected and which waves are absorbed from a red object.

7. Invisible light waves that are a little longer than red and can be felt as heat waves are called what?

8. When light hits a black object, is the light reflected or absorbed?

9. What happens to green wavelengths that hit a blue object?

10. When you stretch out a slinky between two people and shake the end of the slinky from side to side, what kind of wave do you see traveling through the slinky?

11. Give five examples of transverse waves.

12. Can transverse waves travel through space as well as through air?

13. Do waves carry energy?

14. Do different colors of light have different wavelengths?

Pause and Think: Dazzling White and Abysmal Darkness

There are many shades of white, but the purest white is often described as "white as snow." If you look at newly fallen snow while the sun is shining on it, it almost hurts your eyes. People who are out a lot in these conditions generally wear sunglasses. What is happening when you see white snow is that light from the sun hits the snow and practically all the colors of the spectrum are reflected; none of the waves are absorbed. That means all the colors of light enter your eyes when you see white.

There are many references in the Bible to pure white. In Daniel's vision (7:9), he saw the Ancient of Days whose clothing was white as snow. Isaiah (1:18) wrote that sins like scarlet would become white as snow. In another vision of heaven, a multitude of people are seen wearing white robes (Rev. 7:9). What do you think this pure white symbolizes?

Black darkness is the absence of light. When you see white, all the colors of light enter your eyes, but when you see black, none of the colors of light enter your eyes. Black objects absorb the light waves.

There are also many references to darkness in the Bible. One of the most interesting verses is 1 John 1:5, which says God is light and there is no darkness at all in Him. Verse 6 goes on to talk about walking in darkness. What do you think darkness symbolizes?

Discuss: What are some of the things that are symbolized in the Bible by "light" and by "darkness"?

INVESTIGATION #6

Think about This

Do you think each person has a unique voice? Your teacher will assign someone to be "It." This person will sit with his/her back to the others. Each person in the room will take turns saying a two-syllable word or phase, such as "What's up?" from the same location in the room. Each person should speak clearly in their normal voice. See how many times "It" can guess the person who spoke.

If there is time, let students play two notes (same pitch each time) on a musical instrument and see how many people can identify the instrument without looking at what is being played.

Sounds differ in three basic ways — pitch, loudness, and quality. The FBI and other law enforcement agencies are often called upon to help solve crimes by matching voiceprints. They can do this because people have different voiceprints.

WHAT'S UP?

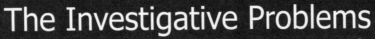

The Investigative Problems
What causes sounds? How do sounds travel from one place to another?
What causes one sound to be different from another sound?

Gather These Things:

✔ Rubber band

✔ Tuning fork (if available)

✔ Box with rubber bands
(from lesson #1)

✔ Metal slinky

Procedure & Observations

Place your fingertips on your neck near where you think your voice box is located and gently exhale. Now gently hum a low note as you exhale. Can you feel a difference when you hum?

Hold one end of a rubber band in your teeth and stretch the band. Pluck the band. Can you tell that it is vibrating? Is there a sound being made by the rubber band?

If you have a tuning fork, your teacher will demonstrate that a vibrating tuning fork will cause water in a glass to splash or make a piece of paper rattle.

Find the box you made for activity #1. You have already noted that the thinnest rubber band produced the highest pitch. Place a block or book on the bottom of the box and slip the rubber bands over it to increase the tension. Pluck the bands again. Did the pitch change?

Hold the metal slinky between you and your partner like you did in the lesson on light. This time you will make a different kind of wave, known as a longitudinal wave. Quickly push forward and then pull back on the slinky with your hand. Place a piece of tape on one of the coils. Describe how the tape moves as you make several longitudinal waves. Recall how a piece of tape moved as you made a transverse wave. Compare the movement of the piece of tape in both longitudinal waves and transverse waves.

Do you see a pulse move from your hand to your partner? Does it bounce back? Try to time your movement so waves going out and waves bouncing back meet in rhythm. Do you see alternating areas where the coils are close together and far apart?

The Science Stuff

All sounds are made by vibrating objects. Sometimes the vibrations can be seen or felt, but many times you don't notice them. Whether or not you can observe that something is vibrating, it is still true that all sounds are produced by vibrating objects.

The **pitch** of a short, tight, or thin rubber band (also a string) is higher than one that is long, loose, or thick. Think about the difference in a bass singer and a high soprano singer. As you would expect, the bass singer would have longer, thicker vocal cords than the soprano.

The lowest pitches are found in long, thick, loose strings and in instruments like big drums or big tubes.

Instruments with a high pitch

flute   violin

Instruments with a low pitch

tuba   guitar

Guitar players can tighten the strings on their instruments to make the pitch of the string get higher.

Sounds can differ in pitch, loudness, and quality. Quality is caused by something called overtones. These are extra vibrations that form. Two musical instruments can hit the same note (same pitch) with the same loudness and still have a different sound because of the extra vibrations.

Slinkies showing compression and expansion

Sound waves are longitudinal waves. Like the waves in the slinky, each sound wave is made up of one compression and one expansion. Instead of sound waves being made up of metal wires, sound waves form in air, water, metals, or other substances.

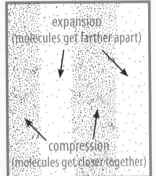

expansion (molecules get farther apart)

compression (molecules get closer together)

A vibrating object pushes and pulls on the molecules around it. Pushing forms a compression, just as pushing the slinky made a compression. Pulling forms an expansion. The closer together one compression is to another, the higher the pitch becomes. Pitch could also be measured by how close one expansion is to another. Loudness is determined by how far each particle moves back and forth. Think about the movement of the piece of tape on the slinky coil.

You observed waves move from one end of the slinky to the other and then bounce back. Sound waves that hit a hard surface and bounce back are known as echoes. When sound waves hit soft surfaces, much of the energy may be absorbed. Sound waves can form and travel through matter. Air, liquids, and most solids allow sound waves to form and move through them. However, unlike light waves, sound waves cannot travel through space.

Making Connections

There is no air on the moon, so there is no sound on the moon. Astronauts who landed on the surface of the moon had radio devices in their helmets. They could send messages to each other by using radio waves, which are transverse waves and can travel through space. However, no matter how loudly they yelled, no one could hear them, because sound waves cannot travel through space.

Have you ever noticed how voices sound in an empty room without carpet, rugs, or curtains? Soft things in a room absorb much of the sound, especially the echoes. Cafeterias and large meeting rooms are usually installed with some type of sound-absorbing materials to prevent echoes and to keep them from being too loud. Many other places contain sound-absorbing materials to prevent sounds from traveling from one room to another.

What Did You Learn?

1. Light waves are what kind of wave?

2. Sound waves are what kind of wave?

3. Why are there no sounds on the moon?

4. Are all sounds made by vibrating objects?

5. Suppose you notice a guitar has a thick string and a thin string that seem to be the same length and have the same tension. Which string would have the highest pitch?

6. What are three ways in which one sound differs from another?

7. Suppose you stretch out a slinky between two people and push and pull one end of the slinky as your partner holds the other end. What kind of wave do you see traveling through the slinky?

8. What tends to be the difference when sound waves hit a hard surface and when they hit a soft surface?

Dig Deeper
Make a musical instrument or a set of instruments that can produce the notes in an octave. Learn to play a song on it and demonstrate your instrument to others.

Learn more about human ears. Draw a picture showing all the parts of the middle and inner ear. Label the parts and give a brief explanation of how an ear receives and transmits sounds to the brain.

What kinds of things must be considered when building large rooms in order to have good acoustics? What happens when a room has bad acoustics?

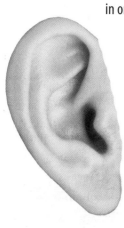

Sound waves travel about 336 meters per second. It takes about five seconds for sound waves to travel one mile. Some airplanes can fly faster than this. Do some research to find out what happens when an airplane goes faster than the sound it makes (supersonic speed).

29

When Things Get Hot

Think about This

Heat is one of the easiest forms of energy to recognize, because you can often feel heat. But if you look up the definition of heat, there is no mention of how it feels. The definition will be something like this: "the energy of moving molecules as the energy is transferred from hotter to colder things." What do you think this means?

We measure how hot or cold things get by using a thermometer. A hot day in Arizona might measure 40°C, while a cold winter day in Alaska might measure −40°C. We will use thermometers in doing some of the investigations today. Your teacher will review how to read the scale and determine the temperatures of various substances.

Procedure & Observations

This activity requires adult supervision. It will help you to better understand the definition of heat. Add one drop of food coloring to a clear glass that is half full of cold water, and add one drop of food coloring to another glass that is half full of hot water. (Your teacher will pour the hot water in your glasses.) Don't stir or disturb the water. Observe what happens over the next few minutes. Then check the glasses every five minutes throughout the rest of the class or until the food coloring has spread evenly throughout the water. Record observations each time you check the glasses.

Wear safety glasses. Before you begin, place some softened candle wax every 4 cm on the metal rod. Have a timer ready so you can record the times. Begin to heat one end of a metal rod in a candle flame as you hold the rod at the other end. Note the time it takes for the heat to travel from the flame to the first piece of candle wax and to each of the next ones. How long does it take the heat to get to your hand?

The Investigative Problems

What observations make us think that particles move faster and farther when heat energy is added?

Gather These Things:

- ✓ Oven mitt
- ✓ Clear drinking straw
- ✓ Two clear glasses
- ✓ Pen or pencil
- ✓ Metal rod
- ✓ Medicine dropper
- ✓ Timer or watch with second hand
- ✓ Red candle wax
- ✓ Modeling clay
- ✓ Weather thermometer (Celsius)
- ✓ Hot and cold water
- ✓ Matches
- ✓ Safety glasses
- ✓ Food coloring
- ✓ Clear glass flask (or bottle with narrow neck)
- ✓ Index card
- ✓ Tape

Remove the rod from the flame as soon as you feel it getting warm. Put on an oven mitt. Hold the rod with the mitt as you continue to heat the rod for another minute. Do you feel the heat now? Blow out the flame and find a place to let the rod cool before you touch it.

Fill a flask or bottle about ⅘ full of water and add a few drops of food coloring to the water. Put the straw in the water and use the clay to hold it in place a few centimeters above the bottom of the flask. Tape an index card to the straw and mark and label the level of the water in the straw on the index card.

Put your hands around the flask to warm it. Notice what happens to the level of the water in the tube. Mark and label this level on the index card. Rub the flask with a piece of ice and mark and label the level of the water after a few minutes. Try to give an explanation for what you see.

Fill a flask or bottle about ⅘ full of water and add a few drops of food coloring to the water. Put the straw in the water and use the clay to hold it in place a few cm above the bottom of the flask. Tape an index card to the straw and mark and label the level of the water in the straw on the index card.

Put your hands around the flask to warm it. Notice what happens to the level of the water in the tube. Mark and label this level on the index card. Rub the flask with a piece of ice and mark and label the level of the water after a few minutes. Try to give an explanation for what you see.

Look at your thermometer and determine the temperature of the air around you. Now predict the temperature of the air near the ceiling (as high as you can reach), near the floor, next to a window, and in one other place of your choice. Record your predictions and then measure the temperature of the air in each place. Allow about one minute for the temperature to change in each place. Record your readings.

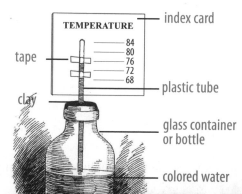

TEMPERATURE — index card

84
80
76
72
68

tape — plastic tube

clay

glass container or bottle

colored water

31

The Science Stuff

Both the molecules of food coloring and the molecules of water are in motion. The food coloring molecules spread out as they bounce off of other molecules and move through the spaces between the water molecules.

The molecules in the hot water have more kinetic energy than the molecules in the cold water. They move faster and farther than the molecules in the cold water. Therefore, the hot water will disperse the food coloring faster than the cold water.

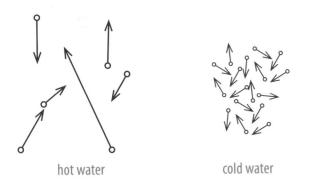

hot water   cold water

The faster-moving molecules have more heat energy than the slower-moving molecules. This is why we can talk about heat in terms of the energy of moving molecules.

Heat can travel easily through metals and other **conductors**. Heat does not travel easily through things that are **insulators**. As the metal rod was heated in a flame, heat energy moved from the flame through the rod to your hand. You could feel the heat with your hand. However, when you held the rod while wearing an insulated oven mitt, you could not feel the heat. The metal rod is a good conductor of heat and easily conducts heat. The oven mitt is made of an insulator, which does not easily conduct heat. The transfer of heat energy through a conductor is known as **conduction**.

Liquids and gases expand when they get hotter and contract when they get cooler. This is because the molecules move faster and through a greater distance as they receive more heat energy. They move slower and not as far as they lose their heat energy.

This is why the liquid and air in the flask expanded when you warmed it with your hands. It is also why the liquid and air contracted when you rubbed ice over the flask. The same thing happens in a thermometer, except the liquid is in a sealed tube.

The temperature of the air in a room will not be the same everywhere. The cooler, heavier air will tend to be near the floor. The warmer, lighter air will tend to rise and be near the ceiling. Currents in the room will help to keep the temperature from being too cold or too hot in places.

The transfer of heat energy through liquids or gases by means of currents is known as **convection**.

Making Connections

Most substances expand when they get hotter and contract when they get cooler. Sidewalks usually have cracks or spaces between blocks of cement. Spaces are left between the metal rails of railroad tracks. Bridges are designed in many ways to allow for hot and cold days. These are a few of the structures where engineers must find ways to keep them from buckling, cracking, or breaking when they expand and contract.

Houses need insulation in the roof and walls. Insulation is often under the floor as well. Insulation keeps the heat inside on a cold day. It keeps the heat outside on a hot day.

Dig Deeper

Do some more research on insulators and conductors. Design something that will keep a glassful of ice water cool for an hour on a hot afternoon, or design something that will keep a cup of hot chocolate hot for an hour on a cold afternoon.

Find out more about what causes convection currents. Draw a picture to show how convection currents move throughout your room.

Do some research on how bridges are built to allow for expansions and contractions due to temperature changes.

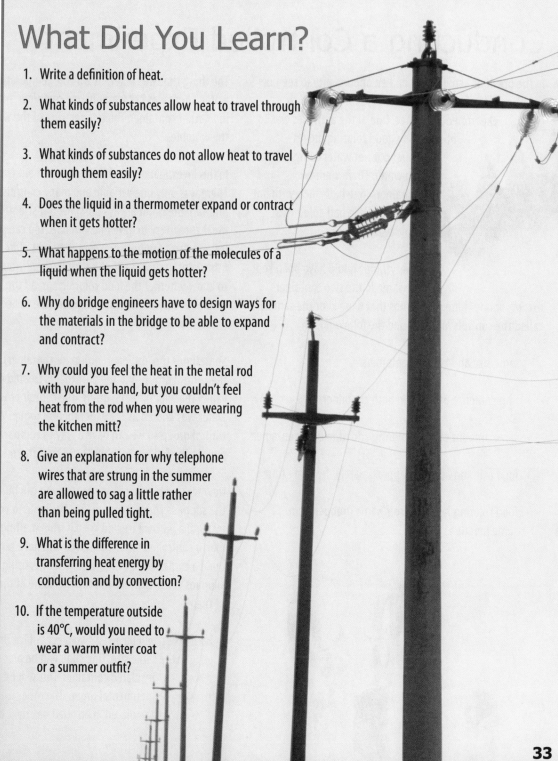

What Did You Learn?

1. Write a definition of heat.

2. What kinds of substances allow heat to travel through them easily?

3. What kinds of substances do not allow heat to travel through them easily?

4. Does the liquid in a thermometer expand or contract when it gets hotter?

5. What happens to the motion of the molecules of a liquid when the liquid gets hotter?

6. Why do bridge engineers have to design ways for the materials in the bridge to be able to expand and contract?

7. Why could you feel the heat in the metal rod with your bare hand, but you couldn't feel heat from the rod when you were wearing the kitchen mitt?

8. Give an explanation for why telephone wires that are strung in the summer are allowed to sag a little rather than being pulled tight.

9. What is the difference in transferring heat energy by conduction and by convection?

10. If the temperature outside is 40°C, would you need to wear a warm winter coat or a summer outfit?

Conducting a Controlled Experiment

In the previous activity, you had an opportunity to see how the principles of controls and variables are applied to experimental science. One of the activities was to observe how food coloring moves throughout hot water compared to cold water. There were two containers, one half-filled with hot water and one half-filled with cold water.

Everything should have been kept the same in the two containers except for one thing. The things that were kept the same are called the controls and included the following:

- same size and kind of containers

- same amount of water in both containers

- same amount of food coloring added to both containers

- food coloring added at approximately the same time

- food coloring released from same dropper from same height

The thing that was different was the temperature of the water. There was hot water in one container and cold water in another container. The temperature of the water was the variable.

In this investigation, a controlled experiment was conducted. There was only one variable and many controls. Having only one variable is not always possible, but scientists will take great care to try to have one variable. You can see why this is important. Suppose one container was almost full and the other container was about half-full. There would be no way to know whether the food coloring spread throughout the water because of the water temperature or because of the amount of water.

Sometimes the "ordinary" group (or container) is called the control group, and the part that is treated differently is called the experimental group. The container with the hot water was known as the experimental group. We need the control group so we can have a way to compare what happens in the hot water to the "ordinary" group.

Now we will have a contest to see who can build a box from the supplies your teacher makes available to you that will insulate a piece of ice and keep it from melting. You must plan a controlled experiment. You will have ten minutes to build a container to insulate an ice cube, using only the supplies you are given. You can use some of them or all of them.

Each group will be given two pieces of ice that are the same size. One cube will be placed in your insulated container. This will be the experimental group. The other ice cube will be placed in an open box next to your insulated container. This will be the control group. Both boxes will be placed in a warm place for 30 minutes. The ice cubes will be measured again. The group with the largest piece of ice in the insulated container is the winner. Be sure to identify the controls and the variable in this experiment.

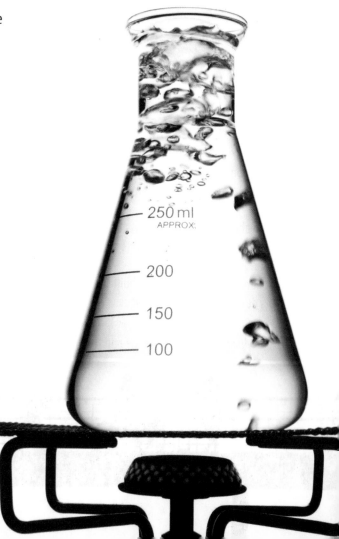

Is all scientific research conducted by doing controlled experiments?

Controlled experiments are a valuable tool for scientists. However, a common misconception is that all scientific research is done by experimental methods using controlled experiments. Some of the other scientific methods include theoretical, correlational, and historical reconstruction. Creative scientists may also come up with many other ways to study the things around us.

This is a good time to meet the Science family, their near relatives, their distant relatives, and the ones who just pretend to be their relatives. The children represent investigative methods used by scientists.

Here are the Science children. They are very curious and ask millions of questions. In their investigations, they look for explanations of things that have been observed in nature, and they look for regular repeating patterns found in nature.

Peri was named after experimental methods. He is the favorite nephew of his aunts and uncle who think he can do no wrong.

Corre was named for correlational studies. She gets a real kick out of finding relationships between things.

Theo was named for theoretical research. He loves a mental challenge and prefers solving puzzles to playing activities with the other children.

Reco was named for historical reconstruction of natural events. The question he asks most often is, "Where did things come from?" instead of "Why do things behave as they do?" Reco knows he wasn't there to observe what happened during the past. He also knows that the further back in time he goes, the more scarce the evidence becomes. Reco does a fantastic job of reconstructing things that are only a few hundred or even a few thousand years old, but he doesn't think it is possible to find a natural explanation for where every living, or once-living, organism came from.

Mame's name came from "many other methods." This child is creative, cooperative, and willing to help wherever she is needed.

Here are near relatives from the Natural History family. They are **Cousin De,** named for descriptive knowledge, and **Cousin Lassi**, named for classification. These children often work with their cousins to help them get started. They observe and describe and classify things in nature, but they don't try to give an explanation for the things they observe.

Another near relative is from the Technology family. **Cousin Tec** is an industrious only child. He sometimes goes by the name of Gine (from a branch of technology called engineering). He doesn't really care why things happen. He is much more interested in how to make them work better, how people can use them, and how they can be used to make money.

The semi-scientists are the distant cousins, **Sem** and **Misci**, who try to act like real scientists, but they take short cuts and don't follow all the rules. They often embarrass the rest of the family by pretending to know more than they really do and getting important ideas mixed up.

The **impostors** try to fool people by pretending to be a part of the Science family. They do their best to sound like the scientists and even use the same words they've heard them say. But, they are fakes. The more you understand about what science is, the better you will be able to recognize semi-scientists and impostor scientists.

Discussion:

Do all scientific investigators use the same methods?
What is the main difference in science and technology?
Is there someone you have seen on TV or in the news who would fit the description of a semi-scientist or an impostor? What would make you think this?
In what ways is Reco different from his brothers and sisters?

Reco's brothers and sisters always look for natural explanations in their research. Why is it difficult for Reco to find an all-natural explanation for where every living or once-living organism came from?

Feeling the Heat

Think about This
Tom came inside the house to escape from the 38°C temperature outside and get a cool drink of water. He noticed the indoor thermometer showed a temperature of 22°C.

There hadn't been even a breeze all morning while he cut the grass, but when he opened the door to go back outside, he felt a breeze. Back outside, the air was perfectly still. Why was there a breeze only in the doorway?

The Investigative Problems
Is heat transferred from warmer things to cooler things?
Is heat transferred through air and liquids by convection currents?

Gather These Things:
✔ Two 12-oz. Styrofoam cups

✔ Hottest tap water

✔ Long alcohol thermometer
(not attached to backing)

✔ One 8-oz. Styrofoam cup

✔ Cold water
(refrigerated or iced)

Procedure & Observations

Before you begin, punch a narrow hole in the bottom of the 8-ounce cup, turn it upside-down, and twist the thermometer into the hole, being very careful not to break the thermometer. The thermometer should extend far enough to go into the water each time. This will make a simple **calorimeter** that will help prevent heat loss and allow you to compare heat gain and heat loss in the hot and cold water.

Fill a 12-oz. cup one-third full with hot water. Fill the other 12-oz. cup one-third full with cold water. Measure the temperature of the water in each cup. (Remove the thermometer if necessary.)

Now, pour the hot water into the cold water to make warm water. Immediately place the calorimeter (inverted cup and thermometer) over the water. Wait one minute for the hot water to mix with the cold water. Take the final temperature of the mixed water.

What is the difference between the temperature of the hot water and the warm temperature (last temperature reading)?

What is the difference between the warm water (last temperature reading) and the temperature of the cold water?

Try to come up with your own explanation for why the hot water got colder and the cold water got hotter. Don't simply say they mixed.

The Science Stuff

Theoretically, the heat gained by the cold water should equal the heat lost from an equal amount of hot water. It probably won't be exactly the same, because the cups will absorb some of the heat.

Heat energy always transfers from warmer objects to cooler ones. When a warm object is in contact with a cold object, the warm object loses heat and becomes cooler, while the cold object gains heat and becomes warmer. They will eventually reach the same temperature. This equalizing of temperature can occur by **conduction**, **convection**, or **radiation**.

When the hot water was poured into the cup of cold water, the molecules with the most energy began to bump into molecules with less energy. The more energetic (hot) molecules transferred some of their energy to the cold molecules. Heat **conduction** occurs as molecules collide and transfer energy.

The other method involved convection currents. The colder water is heavier and sank to the bottom of the container. The warmer water is lighter and rose to the top. Convection currents also form in air, as cool air falls and warm air rises.

The same thing happened when Tom opened the door. Hotter molecules immediately began to bump into the cooler molecules, one at a time. However, the breeze Tom felt was mostly because of convection currents.

Remember that the warmest air in an area is lighter in weight than the cooler air. This caused the lighter, warmer air to rise. The cooler air then moved into the area where the warm air left. This created the breeze.

Making Connections

The sun's energy must pass through millions of miles of space. Heat energy from the sun cannot be transferred to the earth by either conduction or convection, because these processes are not possible in space. Heat from the sun is transferred to the earth by electromagnetic waves through a process known as **radiation.**

Sea breezes and land breezes are common along ocean beaches. During the day, the land absorbs heat from the sun more quickly than the ocean and becomes warmer than the ocean. The warm air above the land rises. The cooler air above the ocean moves in to take its place. This is known as a sea breeze.

At night, the land cools quickly, so the warmer air is above the ocean. As this air rises, the cooler air above the land moves in to take its place. This is known as a land breeze.

Heat is sometimes measured in units called **calories**. This is the amount of heat necessary to raise the temperature of one gram of water one Celsius degree. (Many scientists prefer to do scientific measurements of heat in units called joules.) The unit used to rate the energy in food is actually a kilocalorie (1000 calories), and is often written with a capital C. Everyone needs Calories to have energy and live, but if you eat too many Calories, you may gain unwanted weight.

What Did You Learn?

1. On a cold day, does a well-insulated house keep the cold out or does it keep the heat inside?

2. Give three ways in which heat can be transferred from one object to another.

3. Explain what causes a sea breeze. Do they occur during the day or at night?

4. What are two units used to measure heat energy?

5. How does the sun's energy reach the earth?

6. Does warm air tend to rise or sink?

7. When a hot object is next to a cold object, how does heat always move?

Dig Deeper
Find the average number of Calories a person your age needs each day. There may be a difference in what is needed by males and females. Make a list of all the foods you ate yesterday and estimate the total number of Calories you took in. At this rate, will you gain, lose, or maintain weight?

Some materials have the ability to store heat longer than others. We have already noticed that on a beach water can store heat longer than land. This property of matter is known as specific heat capacity. Explain the meaning of this term and give the specific heat capacity of several substances, including water. How does this property of water affect living things?

Explain why insulating your house will help to reduce your heating bill. Why will insulating your house help to reduce your cooling bills?

Magnets Are Very Attractive

Think about This

Juan thought his magnet was a fake, because it wouldn't pick up the pennies he had dropped. He was about the throw it away, when his brother told him magnets aren't supposed to pick up pennies. They are made of copper, and a magnet will not pick up copper.

"Guess what?" his brother said. "The magnetism will go right through a penny."

We will investigate some of the things a magnet will attract, some of the things a magnet will not attract, and some things that magnetism will go through. Do you already have some ideas about this?

Procedure & Observations

Touch the magnet to each item in the tray, one at a time. Place the items that are attracted to the magnet in one stack. Place the items that are not attracted to the magnet in another stack. Fill in the chart below.

Things a magnet attracts	Things a magnet does not attract

Are all metals attracted to a magnet?

Are there any nonmetals that were attracted to a magnet?

Now we will test these same items to see which items will allow a magnetic field to pass through them. Assemble this equipment: Stack books and place a ruler between the top book and the next one. Tie a horseshoe magnet to the end of the ruler with the poles hanging down. (Alternatively, tie another kind of magnet to the ruler.) Tie a string to one end of a paper clip and tape it to the table. Make certain that the paper clip is drawn upward toward one pole of the magnet. Leave a small space between the paper clip and the magnet.

stack of books — — ruler
— horseshoe magnet
— paper clip
— string
— tape

The Investigative Problems

What are some things that magnets attract and some things they don't attract?
Do electromagnets behave like other magnets?

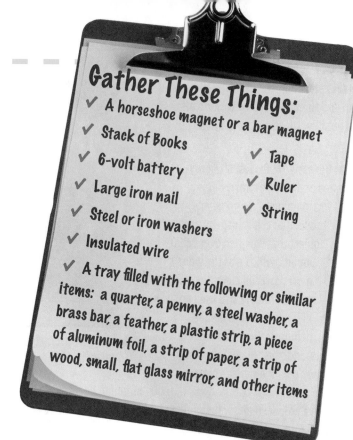

Gather These Things:

✓ A horseshoe magnet or a bar magnet

✓ Stack of Books

✓ 6-volt battery

✓ Large iron nail

✓ Steel or iron washers

✓ Insulated wire

✓ Tape

✓ Ruler

✓ String

✓ A tray filled with the following or similar items: a quarter, a penny, a steel washer, a brass bar, a feather, a plastic strip, a piece of aluminum foil, a strip of paper, a strip of wood, small, flat glass mirror, and other items

Pass the various items on the tray between the magnet and the paper clip. Note which materials allow the magnetic field to pass through and which ones do not. To get accurate results, you will have to be sure that the object doesn't touch either the magnet or the paper clip.

Record in the chart below.

Things that pass through a magnetic field	Things that do not pass through a magnetic field

According to your observations, which materials would allow the magnetic field to pass through them?

Which did not?

Compare the two charts. How are they alike?

An electromagnet has been assembled by the teacher. See if it will attract the same materials the other magnet did. You will have to work fast on this. The wires in an electromagnet get hot and the battery will go dead in a short time. It is a good idea to disconnect one of the wires in between tests.

Try to pick up several iron washers with the electromagnet. While they are suspended, disconnect one of the wires to the battery. What happens?

ELECTROMAGNETIC POWER

The Science Stuff

Magnets will pick up or attract items that are made of iron, steel, nickel, and cobalt. These materials are known as magnetic materials. Nickel coins are not usually attracted to a magnet, because they have such a small amount of nickel metal in them.

Magnets will not attract items made of nonmagnetic materials, such as aluminum, copper, paper, plastic, or glass.

Magnets have a magnetic field around them. Nonmagnetic materials will be able to pass between the magnet and the paper clip without affecting the magnetic field. Normally, when a magnetic material, such as an iron strip, is passed through the gap, the magnetic field will not pass through it, which means that the paper clip will fall.

An electromagnet will attract the same materials that a regular magnet will attract. Regular magnets and electromagnets are similar in this and other ways. They both have north and south poles and they both have a magnetic field around them.

Electromagnets are different from regular magnets in that they can be turned on and off. Another difference is that their poles (north and south) will switch if the wires to the battery are reversed.

Making Connections

Lodestone is a natural magnet. It is a compound made of iron and oxygen. Most forms of iron oxide are known as iron rust, which is neither a magnet nor a magnetic material. Lodestone's particular crystalline shape is thought to be the reason for its magnetic properties. The first lodestone rocks were found in a region near Greece known as Magnesia. The word *magnet* is taken from the word *Magnesia*.

One place you should never stick a magnet is on a TV or computer screen. The magnetic field will cause some of the materials in the screen to move toward the magnet, and the picture will be distorted.

Sometimes carpenters work with magnetized screwdrivers. This lets them hold the screw in place to start with. It also helps them pick up screws that may accidentally be dropped.

Another neat trick for finding screws that fall and roll behind furniture is to tie a string to a magnet and slide it around until the screws are attracted to it. This is also a good way to find nails that have fallen on the ground after a building project. Nails that are not picked up often wind up stuck in car tires. This will not work for lost pennies or for things made from aluminum or tin, because they are not magnetic materials.

Dig Deeper

Powerful electromagnets can be used to pick up heavy steel objects, like cars. The cars can be picked up, moved, and then released by turning off the electric current. Find out how these electromagnets are made to be so powerful.

During the early 1800s, Joseph Henry tested and built several models of electromagnets. He made one model that could suspend an object that weighed over 2,000 pounds, which was the world's most powerful electromagnet at the time. He also used an electromagnet to build a demonstration model of the world's first telegraph. Henry was one of America's most influential scientists during the 1800s, although he often did not receive proper credit for his work. See if you can find out more about his contributions to the field of science.

Photo of an electromagnet taken in 1911

Photo showing the attraction of paper clips to magnetic lodestone

What Did You Learn?

1. Will a magnet pick up a copper penny?

2. Will a magnet pick up an iron nail?

3. If you put a sheet of aluminum under a magnet, will this prevent the magnetic force from going through it?

4. Which of the following materials are magnetic materials: glass, steel, nickel, wood, water, gold, cobalt, magnesium, oxygen, iron, plastic?

5. Which of the materials listed in #4 could a magnetic field pass through?

6. How are regular magnets and electromagnets alike?

7. How are regular magnets and electromagnets different?

8. What is lodestone?

In a horseshoe magnet, the north and south poles are located close together, creating a strong magnetic field between them. There will be more nails attracted to the area near the poles than to the middle of the magnet.

INVESTIGATION #10

Think about This
Tie a string around a bar magnet and suspend it from some place you can easily see. The teacher will have several students, one at a time, slowly walk toward the magnet from across the room in an east-west direction. Two students will have magnets hidden in their pockets. One of the magnets will be stronger than the other one. The suspended magnet will begin to turn when another magnet gets close enough. See if you can figure out which students are hiding magnets and which student has the strongest magnet. Something is obviously happening in the space around the magnets, because the magnets are not touching each other. What is causing the suspended magnet to turn? Which student had the hidden magnet that caused the suspended magnet to turn?

The Investigative Problems
What special properties does a magnet have? In which direction does a bar magnet that is suspended on a string point when it stops moving? What is a compass?

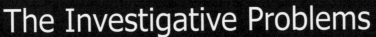

Gather These Things:
- ✓ Two bar magnets
- ✓ Ruler
- ✓ Compass
- ✓ Steel pin
- ✓ Thread
- ✓ Matchbox
  (or other magnet holder)
- ✓ Tape
- ✓ Speaker magnet or other strong magnet

Procedure & Observations

Place a bar magnet in the outer part of a matchbox. Use tape to keep the magnet from sliding out the end of the holder. Push a piece of string through the holder and tie the string to a stationary object. The magnet should swing freely.

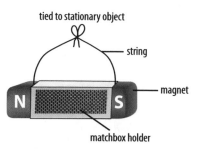

tied to stationary object

string

magnet

matchbox holder

Bring the S (south) end of another magnet toward the S end of the hanging magnet. Bring the N (north) end toward the N end of the hanging magnet. Now bring the N end of the magnet toward the S end of the hanging magnet. This time bring the N end of the magnet toward the S end of the hanging magnet. Record what happens each time.

Cause the hanging magnet to stop moving. Spin it slightly and allow it to come to a complete stop on its own. Repeat several times and record which way the magnet points when it stops.

Place a compass on the table in front of you. Make sure it is not near any of the magnets. Turn the compass around once and wait for the needle to stop moving. Compare the direction the needle in the compass points with the direction of the hanging magnet when it stops moving.

Recall the demonstration you saw at the beginning of the class. Use these same two magnets for this investigation. One magnet should be stronger than the other. Walk slowly toward the hanging magnet as you hold the first magnet. Measure the distance at which you first see the hanging magnet start to move. Repeat, using the second magnet. Do both magnets affect the hanging magnet from the same distance?

Lay a centimeter ruler on the table and place a pin at the end of the ruler. Slowly slide a magnet from the other end of the ruler and note when the pin jumps to it. Record the distance between the magnet and the pin when the pin first moves. Repeat the activity with another magnet. The magnet that makes the pin jump to it from the greatest distance is the stronger magnet. Determine which of the magnets is the most powerful (has the strongest magnetic force) by how far away it is from the pin when the pin moves.

Predict what will happen when you place the N end of one magnet near the center of the other magnet. Try it and see what happens.

You probably can tell just from feeling the effects that the magnetic force is strongest at the poles of a magnet. Try to push two like poles closer together with your hands. Record how it feels. Now try to hold two opposite poles apart with a small space between the poles. Record how it feels.

The Science Stuff

The N pole of one magnet is attracted to the S pole of another magnet. Two N poles will repel each other. Two S poles will repel each other. The **law of magnetic attraction** states that two like poles repel and two opposite poles attract. This ability to attract and repel is the special property used to identify magnets.

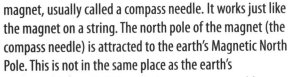

The attracting and repelling forces that are observed between two magnets occur even when the magnets are not touching. The space around a magnet, where the force of a magnet works, is known as the **magnetic field**.

What we call the **north pole** of a bar magnet is, in reality, the **north-seeking pole** of the magnet. You have learned that the south pole is attracted to the north pole. If a magnet is suspended by a string, the end that is attracted to the **Magnetic North Pole** of the earth will turn until it points north. You would logically conclude that the south pole of the magnet would be attracted to the North Pole. But over time and by tradition, the end of the magnet that points to the North Pole of the earth is called the north-seeking end of the magnet. This north-seeking pole eventually came to be labeled the N pole.

A compass is a case that contains a freely turning magnet, usually called a compass needle. It works just like the magnet on a string. The north pole of the magnet (the compass needle) is attracted to the earth's Magnetic North Pole. This is not in the same place as the earth's **Geographic North Pole**, which you see on a world map. However, the two poles are close enough so that the direction in which a compass needle points is probably almost correct.

When one of the poles of the bar magnet is placed near the center of the other magnet, there will be a weak attraction or repulsion. The attraction or repulsion between two magnets is strongest at the poles.

You can determine which of several magnets is the most powerful (has the strongest magnetic force) by a simple test. The one that can cause a magnetic material to move from the greatest distance is the strongest.

Making Connections

Early sailors made compasses by putting a piece of lodestone in a case and floating it in a small amount of water so it could turn freely. The north-seeking end of the lodestone pointed north. Compasses became more refined until they look much like today's models. They were valuable pieces of equipment for people traveling by ships. The sun by day and the North Star by night helped them stay on course, but if the sky became too cloudy, their compass was the only other way to know which way they were going.

Compasses are still valuable tools today. They are often used by airplane pilots, campers, mountain climbers, hikers, and others to help determine their location or to keep from getting lost. Many cars now have built-in compasses in their mirrors.

Dig Deeper

The north pole of a freely turning magnet doesn't always point true north. It points to the earth's Magnetic North Pole, which is about 1,100 miles from the earth's Geographic North Pole. Draw (or obtain) a map showing the location of each pole. Show the location of your hometown or city on the map. Draw a straight line from your home to the Magnetic North Pole. Would a freely turning magnet point true north from your home? Mark some other cities on your map and connect each city to the Magnetic North Pole. How would a compass point from each of these cities?

To use a compass accurately, one must understand what is meant by magnetic declination. Explain what is meant by magnetic declination and tell how pilots, hikers, campers, and others use declination charts to help determine the direction in which they are traveling. Hint: Scouting manuals provide good information and instructions about using a compass.

Do some research on how early sailors made and used compasses. When and where were some of the first compasses thought to be made?

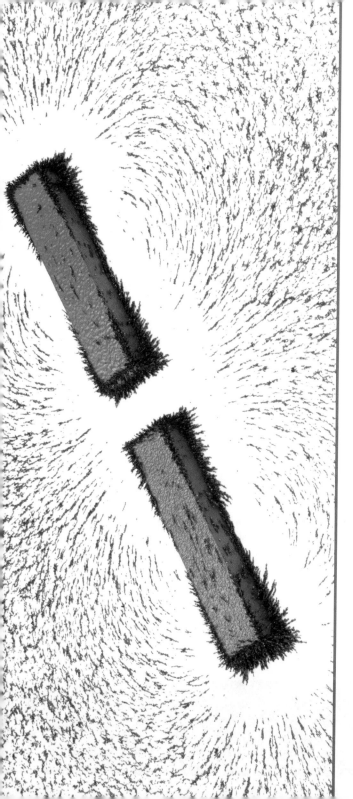

What Did You Learn?

1. Under what conditions will two bar magnets push away from each other, even when they are not touching?

2. Under what conditions will two bar magnets pull on each other, even when they are not touching?

3. In what direction does a bar magnet that is suspended on a string point when it stops moving?

4. In what direction does a compass needle point when it stops moving?

5. Which end of a freely turning magnet points north?

6. What is one way to compare the strength of magnets?

7. Is the north pole of a freely turning magnet attracted to the earth's Geographic North Pole or the Magnetic North Pole?

How Do Magnets Become Magnets?

Think about This
Sam had a set of magnets that came with a science kit. He let his younger brother, Max, pick up paper clips with them. Max soon got tired of the paper clips and began to pretend that the magnets were hammers and bombs. One day, Sam noticed that the magnets could no longer pick up paper clips. What did Max do to his magnets?

The Investigative Problems
What happens when an object becomes magnetized?
How can objects that have been magnetized lose their magnetism?

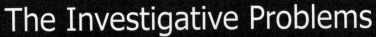

Procedure & Observations

Put iron filings into the tube and lay it flat on the table. Slide the compass next to the tube from one end of the tube to the other. Observe what the compass needle does.

Now, with a strong magnet rub along the side of the tube 30 times in one direction, always rubbing with the same end of the magnet. Move the compass along the side of the tube again. What did the compass do after the tube had been rubbed with a strong magnet?

Shake the tube several times and move the compass along the side of the tube. Did the compass behave the same way it did the last time?

Remove the iron filings and replace them with small cut-up pieces of aluminum. Repeat the experiment the same way you did with the tube of iron filings. Tell how the compass behaved each time.

Take a large steel paper clip and straighten it. Touch it to a small steel paper clip. What happens? Rub the straightened clip several times in the same direction with a magnet as you did earlier with the tube of iron filings. Now touch the straightened clip to some small paper clips. What happens? Tap the straightened clip on a hard surface several times. Touch it to the small paper clips again. What happens?

The Science Stuff

This activity shows how a magnetic material can become magnetized. Magnetic materials are usually made of iron (steel), cobalt, and nickel. They contain tiny magnetized regions known as domains. These domains have north and south poles. When the domains are arranged in a random manner, the magnetic material is not a magnet. When many of the domains are lined up in the same direction, the magnetic material becomes a magnet with its own magnetic field.

Nonmagnetic materials do not contain magnetic domains. They are not attracted to magnets, and they cannot be made into magnets. Nonmagnetic materials generally do not contain the elements iron, nickel, or cobalt.

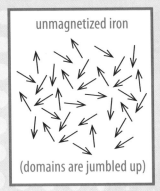

unmagnetized iron

(domains are jumbled up)

magnetized iron

(most domains are lined up in the same direction)

The iron filings in the tube behave like magnetic domains. At first, the domains are mixed up in a random way and the tube has no magnetic poles. When a strong magnetic field keeps pulling on the domains in the same direction, many of them begin to turn and line up the same way.

Compare that to the diagram that shows how these tiny magnets look when they are lined up in an organized way in a magnet.

When many of the domains are lined up in the same way, the tube of filings will have a north pole and a south pole. It will also have a magnetic field around it.

Shaking the tube will once again mix up the domains, and the tube will lose its poles and its magnetic field. Shaking the tube and randomizing the domains is like what happens to magnets that lose their strength. If a magnet is dropped, hit, or heated, some of the domains will get out of alignment, and the magnet will lose some of its strength. When a magnet has been dropped or hit repeatedly, it may become just an iron bar with no magnetic properties.

If the tube is replaced with small pieces of aluminum, it will not acquire magnetic properties. Aluminum is a nonmagnetic material and does not have magnetic domains that can line up in the same way iron filings did.

Recall that the compass you used in this activity contains a needle that is a magnet. The magnet in the compass can be repelled or attracted by another magnet that is nearby.

Making Connections

Proper storage of magnets will increase their life. Bar magnets should be stored in pairs, with opposite poles next to each other. A non-magnetic material such as a Popsicle stick should separate them.

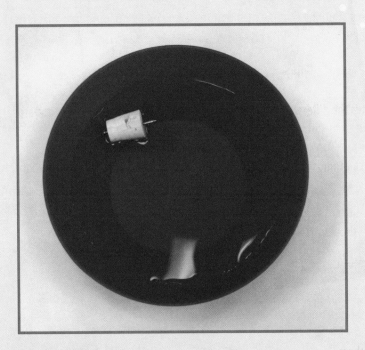

Dig Deeper

Learn how to make your own compass. Try these methods and see if you can make them work:

1. Take a thin iron nail and rub it several times in the same direction with a permanent magnet. Push the nail through a cork and let it float on a pan of water. The magnetized nail should turn in a north-south direction when it stops moving.

2. Place a rod-shaped piece of iron on a firm surface and point it toward the north. Tap it several times, giving the earth's magnetic field a chance to turn a few of the magnetic domains and make them line up in the same direction. Tie a string around the piece of iron so that it is balanced and can turn freely. Mark the end that was pointing north when it was tapped. If you were able to turn some of the magnetic domains, the piece of iron will point north from different locations.

What Did You Learn?

1. What do magnetic materials have in them that other materials do not have?

2. Name something that is a magnetic material and name something that is not a magnetic material.

3. Can a bar of copper be made into a magnet? Why or why not?

4. What is the difference in how the domains are arranged in a magnetic material that is not magnetized and in a magnet? (A diagram is a good way to answer this.)

5. Why can things like wood, lead, tin, or plastic materials not be made into magnets?

6. Can a magnetic material that is not magnetized become magnetized? Explain.

7. How might you cause a magnet to lose its strength?

8. Why would a manufacturer of magnets not guarantee them if you drop them?

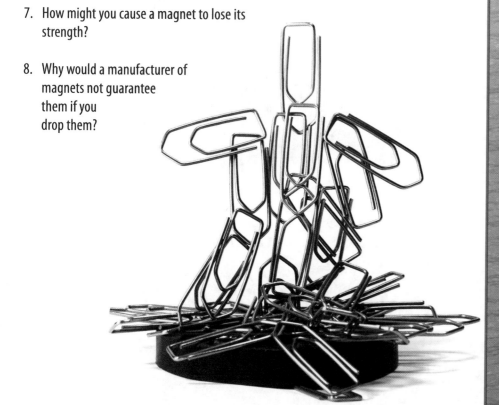

Writing Project: The Rescue

Try to write an adventure story about you and an older friend. Begin your story about a camping/canoeing trip. Tell about getting caught in a sudden thunderstorm several miles downstream where the canoe crashes, most of the equipment gets lost in the river, and your friend is injured. The only thing you can recover is a fishing tackle box that has fishhooks, some iron nails, a knife, fishing line, and a flashlight. With only a few minutes of daylight left, the sky is too cloudy to find the North Star. The compass was lost, but you need to be able to walk toward the highway directly north of your present location. Be creative and tell how you find help and rescue your friend.

If It's Invisible, How Can You See It?

Think about This
We have discovered that magnets have an invisible force that attracts a piece of iron even when they are not touching. Do you think there is any way to see what this invisible force looks like?

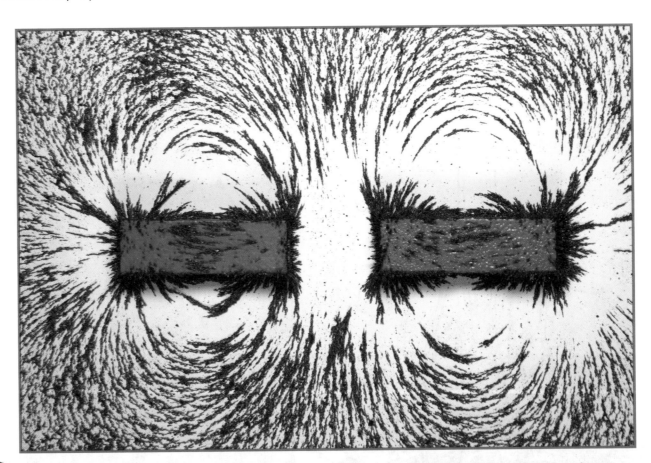

The Investigative Problems
How can we see what a magnetic field looks like?
Where are magnetic lines of force?

Gather These Things:

- ✔ Two bar magnets
- ✔ Compass
- ✔ Large piece of paper or cardboard
- ✔ Shaker of iron filings
- ✔ Two thin pieces of wood about the thickness of the magnets
- ✔ Long piece of insulated wire
- ✔ Large iron nail
- ✔ Battery
- ✔ Covered thin box

Procedure & Observations

1. Place the bar magnet midway between the two pieces of wood. Place the paper over the magnet. Carefully sprinkle the iron filings on the paper, over and around where the magnet rests. Very gently tap the edge of the paper and notice what the iron filings do.

2. Make a drawing of the pattern you observe. Hold a compass near the north pole and the south pole of the magnet, and note how it turns in each position. Observe how it turns from various other places. When you finish making your observations, use the paper to pick up the iron filings and return them to the shaker.

3. Place two magnets about two centimeters apart with north poles facing. Place them between the two pieces of wood and cover them with a piece of paper. Sprinkle iron filings over the paper. Very gently tap the edge of the paper and make a drawing of the pattern you observe. When you finish making your observations, use the paper to return the iron filings to the shaker.

4. This time, place the opposite poles about two centimeters apart and sprinkle filings over the paper. Tap the paper, make a drawing of the pattern you observe, and note how the compass turns. Return the iron filings to the shaker when you finish.

5. Work with a partner. Place a magnet in a thin, covered box and tape it in place. Using only a compass, see if your partner can accurately identify where the magnet is located, as well as which end is the N pole and which is the S pole. Fold a piece of paper to be the size of the box top and draw how you think the magnet is placed. When you finish, open the box to see if you made an accurate drawing.

6. An electromagnet has been assembled by the teacher, and iron filings have been sprinkled over it in the same way the first activities were conducted. Tap the paper gently until you can see a pattern. Does it look like the first drawing you did? How does a compass behave near it? Does this magnetic field look like the one you observed around a permanent magnet? Disconnect one of the wires to the battery as soon as you see the pattern of the magnetic field.

The Science Stuff

The magnetic field around a magnet is made up of invisible **lines of force** that connect a magnetic north pole and a magnetic south pole.

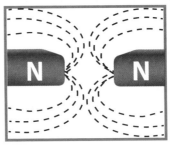

Notice that the earth has the same kind of magnetic field pattern around it as a bar magnet does. The earth also has a north pole, a south pole, and magnetic lines of force.

Magnetic North Pole

Axis

Magnetic South Pole

The lines of force can form from the poles of the same magnet or they may form from different magnets. It may look like the lines of force extend a little way from the magnet and stop, but the paper may not be large enough for all of them to show up. Lines of force always connect a north pole and a south pole.

The iron filings align themselves along the lines of force within the magnetic field. This lets us see the pattern made by these invisible lines of force. Notice the filings tend to cluster at the poles where the lines of force are very close together. Recall that the poles are the strongest parts of a magnet.

Recall that all magnets have two poles. Two like poles repel each other and two unlike poles attract each other. When two north ends or two south poles are placed together, the lines of force are pushed away from each other. When opposite ends are placed together, strong lines of attraction form between the poles.

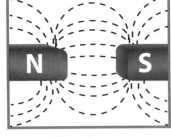

like poles repel

unlike poles attract

Recall that an electromagnet behaves like a permanent magnet, except that it can be turned on and off. Electromagnets have poles and a magnetic field as long as an electric current is flowing through it.

Actually, a straight wire that has a moving current of electricity through it will have a weak magnetic field around it. An electromagnet can be made by wrapping a long electric wire around an iron nail several times to concentrate the field. The ends of the wire are connected to the + and − posts of a battery. As long as the current is flowing, the coiled wire will act like a bar magnet. When the current stops, the magnetism will go away. Switching the wire connections to the battery will reverse the north and south poles of the electromagnet.

Making Connections

The earth acts like it has a giant magnet inside it. The earth has a north pole (that is actually in the south) and a south pole (that is actually in the north). Magnetic lines of force connect the earth's north and south poles.

However, scientists believe the earth's magnetism is more like an electromagnet than a bar magnet. Scientists believe the earth's core is made of iron, a magnetic material, but, as we learned, heating is one of the ways the domains in a magnetic material get jumbled up. Although it is very hot, they believe the movement of the molten currents produces a large electric current. As long as there is a moving electric current, there will be a magnetic field around it.

Have you ever seen or heard about the northern and southern lights? The earth's magnetic field deflects many solar radiations (some quite dangerous) away from the center of the earth. The waves and particles follow the earth's lines of force toward the poles. As they approach the earth's two magnetic poles, they pass through the earth's atmosphere in large numbers and create a beautiful display of lights. If you've never seen these lights before, you can see what they look like by doing a search for "northern lights" on the Internet.

Dig Deeper

1. Scientists have been measuring the strength of the earth's magnetic field for about 150 years and have found that the magnetic field is continually getting weaker. If it continues to decay at this rate, it may lose about half its strength in another 1,400 years.

   Dr. Thomas Barnes believes the weakening of the earth's magnetic field is evidence for a young earth. If time could be run backward, after only several thousand years, the magnetic field would be too strong for life to exist. As time continues, the magnetic field seems to be winding down and gradually getting weaker.

   Look for more information about the theory that the earth's magnetic field is gradually getting weaker.

2. There are a number of theories about how migratory birds and animals are able to navigate their routes. One theory is that they have an ability to detect the earth's magnetic field. Homing pigeons, sparrows, and whales are a few of the animals that may have senses that can detect the earth's magnetic field. Choose a migratory bird or other animal that may have special senses for navigating by means of the earth's magnetic field. Show this animal's migratory route on a map. Find some other information about this animal. If the earth's magnetic field became much weaker or even disappeared, how might that affect migratory birds and animals?

3. Would a weaker magnetic field allow more of the dangerous radiations from the sun and outer space to enter the earth's atmosphere? Does our moon have a magnetic field? Would there be a problem from solar radiations if astronauts tried to build a colony on the moon? Explain.

What Did You Learn?

1. What is a compass needle?

2. On what did the iron filings become aligned when you sprinkled them over a magnet?

3. A magnetic line of force extends from what to what?

4. Does the earth have magnetic lines of force that go from the earth's magnetic north pole to the earth's magnetic south pole?

5. Explain how the earth's magnetic field protects us from many of the harmful radiations that come from the sun.

6. Where are the magnetic lines of force around a magnet closest together?

7. What do scientists believe causes the earth to have a magnetic field?

8. How would you describe the interaction of magnetic poles?

9. Is there a magnetic field around a magnet?

10. Is there a magnetic field around a moving current of electricity?

Static Electricity

Think about This "Okay, Mom, I'm getting the clothes out of the dryer," Jonathan yelled as he walked into the laundry room.

"What do you know — the light is burned out and this room is pitch dark," he muttered. No problem. He'd done this plenty of times before and knew exactly where the dryer was.

"Wow! Would you look at this," he said as he pulled the clothes out of the dryer. There were flashes of light and popping and crackling sounds coming from the clothes.

What do you think was going on?

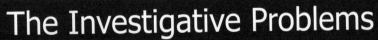

The Investigative Problems
How do some objects acquire a static charge?
How do objects with a static charge interact with neutral objects?

Gather These Things:

- ✓ Two plastic strips
- ✓ Hard rubber comb
- ✓ Two balloons
- ✓ Wool cloth
- ✓ Puffed rice cereal
- ✓ Narrow glass jar
- ✓ Nylon cloth
  (or plastic bag)
- ✓ Paper
- ✓ String

Procedure & Observations

1. Pick up the two strips of plastic by one end and hold them close together but not touching. Now lay the strips on the paper and rub them with the wool cloth. Pick up the strips again, being careful to touch only one end. While they are hanging down, slowly bring them together. Record what you observed.

2. Vigorously rub the comb with the wool cloth. Slowly bring the comb close to the rice cereal, but don't touch it. Hold it steady. Be patient. Sometimes it takes a few minutes for anything to happen. Record your observations. Repeat the investigation with a blown-up rubber balloon. What happens?

   If you'd like to do a couple of "just for fun" activities, put the rice cereal in a small cardboard box and stretch a sheet of plastic wrap over the box. Rub the plastic wrap rapidly with a wool cloth. OR, charge the comb again and hold it next to the hair of someone who isn't wearing hair spray or an oily dressing. The explanations are the same as for the comb (or balloon) and cereal.

3. Take two inflated balloons and tie a piece of string to each one. Hang the balloons by the strings so they are about two or three centimeters apart. What did the balloons do? Rub both balloons with the wool cloth and observe what happens. Now, rub a narrow glass jar with a nylon cloth or a plastic bag and bring the jar near each of the balloons, one at a time. Record your observations about what happened to the balloons in each case.

The Science Stuff

The three experiments you just did are all explained by static electricity. But what is static electricity? In order to understand what this is, you need to review what you have learned about atoms.

Everything in the world is made up of atoms. Let's imagine that you take an iron nail and cut it in half. Then take one of those halves and cut it in half. Keep taking a half and cutting it in half again until you have a tiny, microscopic piece of iron. Imagine cutting it until you have only one **atom** of iron left. This is the smallest piece of iron you can have.

Believe it or not, you can cut up the atom into smaller parts, but it wouldn't be iron anymore. It would be **electrons**, **protons**, and **neutrons**.

An atom has a nucleus (center) made up of protons and neutrons. The protons have a positive charge and the neutrons are neutral, which means they have no charge. Swirling around the nucleus are tiny electrons, which have a negative charge.

It's very simple: Protons are in the nucleus and are positive (+).

Neutrons are in the nucleus and are neutral (no charge).

Electrons are outside the nucleus and are negative (-)

Because the electrons are outside the nucleus whizzing around, they can sometimes come loose and attach to something else.

This happens sometimes when one material is rubbed with a different material. If an atom loses an electron, it then has an overall positive charge. The atom that gains an electron has an overall negative charge.

When you rubbed a substance made of rubber with a wool cloth, the rubber gained electrons and became negative. The wool lost electrons and became positive. In the process, you created static electricity.

Remember what you learned about magnets? Opposites attract and like things repel. The same thing applies to the positive and negative charges of static electricity. One negatively charged object will attract a positively charged item, or even a neutral item. Two things that have the same charge will repel each other.

When a neutral object is brought close to a charged object (+ or -), the charged object will pull some of the loosely held electrons closer to the surface or push some of the electrons away from the surface. This can be done because a charged object has an electric field around it, much like a magnet has a magnetic field around it. This is what happened to the puffed rice. The rubber comb or balloon became negatively charged when it was rubbed with wool. It pushed some of the electrons down into the cereal, leaving the surfaces with a slightly positive charge. Therefore, the charged object will attract the object that used to be neutral. This why the puffed rice was attracted to the comb.

A similar thing happened to the two balloons. Starting off, they were both neutral. When they were rubbed with a wool cloth, they both took on a negative charge. Two negatively charged balloons repelled each other (pushed apart).

When the glass was rubbed with nylon, it became slightly positive. The negative balloons were attracted to the positive charge in the glass.

Don't be confused by what you learned in the previous lesson. You learned that there is a magnetic field around a moving current of electricity. Now you learn that there is an electric field around an object with a static electric charge. This was also confusing to early scientists who were looking for a way to produce electricity from magnetism. But because someone finally learned the secret of producing electricity from magnetism, we have electricity in our houses and buildings today.

Making Connections

There are a number of products that can help you prevent or get rid of static electricity in your clothes. Some products are put directly in the dryer, and others are sprayed on your clothes after they are dry.

If you comb your clean, dry hair on a day when the humidity is low, you may notice your hair is beginning to stand out all over your head. This is because the combing process gave each hair the same kind of charge. Since like charges repel each other, one hair repelled another, causing them to get as far apart as possible.

Dig Deeper

Charge a balloon with a wool cloth and make it stick to the wall. Rub a half sheet of newspaper with a wool cloth and stick it to the wall. Use a watch to keep up with the time and see how long each will remain stuck to the wall. Repeat this activity, but this time, spray an antistatic product on the balloon and the paper after about one minute. How long will they continue to stick to the wall? Find some information about how antistatic products work.

Make up your own static electricity experiment (or find directions in a reference book or the Internet). Write directions for how to do the experiment and write a scientific explanation for what happens.

What Did You Learn?

1. Suppose you feel a little shock after walking across a carpeted floor and touching a metal door handle. Is this kind of shock caused by current electricity or static electricity?

2. Did protons, neutrons, or electrons move between your hand and the metal door handle to cause the shock?

3. Atoms contain positive, negative, and neutral particles. Which two kinds of particles are found in the nucleus of an atom?

4. What do you call the space around an object with a positive or negative static charge?

5. What kind of field is around a moving electric current?

6. Do electrons have a positive or a negative charge?

7. What happens when two negatively charged balloons come near each other?

8. What happens when a positively charged balloon comes near a negatively charged glass?

9. What happens when a negatively charged comb comes near a neutral piece of rice cereal?

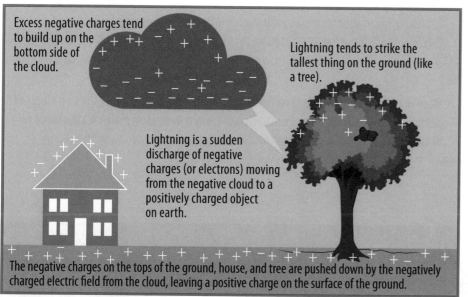

Excess negative charges tend to build up on the bottom side of the cloud.

Lightning tends to strike the tallest thing on the ground (like a tree).

Lightning is a sudden discharge of negative charges (or electrons) moving from the negative cloud to a positively charged object on earth.

The negative charges on the tops of the ground, house, and tree are pushed down by the negatively charged electric field from the cloud, leaving a positive charge on the surface of the ground.

Pause and Think: Lightning Safety

Did you know that lightning is a huge discharge of electrons that were built up from static electricity? A negatively charged cloud can do the same thing to the earth as a negatively charged comb over puffed rice cereal. The negative electric field under the cloud will push some of the electrons in the ground down deeper. The top of the ground will then have a positive charge. Eventually, a large number of electrons from the cloud will jump (be discharged) from the cloud to the earth as a bolt of lightning.

Did you know that a bolt of lightning can travel 60,000 m/sec (135,000 mi/hr) and reach a temperature of 30,000°C (54,000°F), which is enough to fuse soil or sand into glass? There are 60 million thunderstorms every year all around the world.

Remember that lightning can travel easily through metal and other conductors. Lightning can travel through water. Lightning tends to strike the tallest things around.

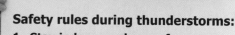

Safety rules during thunderstorms:

1. **Stay indoors and away from open windows and doors.**
2. **Stay out of bathtubs and showers.**
3. **Do not touch water faucets, metal pipes, or plug-in appliances such as an iron.**
4. **If you are outside, stay away from tall trees or poles or power lines.**
5. **If you are in an open field, you are the tallest thing around. Lie down if you feel your hair starting to stand on end. This means you are being charged.**
6. **If you are in a boat, get to land.**
7. **Never swim during a thunderstorm.**
8. **Do not carry metal rods outdoors in a storm.**
9. **If you are in a car, stay there. Charges will be conducted to the ground.**

Discussion: Where are some places you should not be during a thunderstorm?

A Place Where Electrons Get Pushed Around

Think about This

It would be fun to be able to detect the presence of an electrostatic charge on various objects. With the following procedure you can make a simple but effective electroscope, which can do just that.

Hans Christian Oersted (1777 – 1851) was a Danish physicist and chemist. He is best known for discovering that electric currents can create magnetic fields.

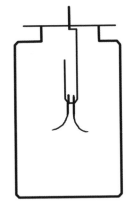

Procedure & Observations

First, make a simple electroscope from these instructions.

1. Straighten a paper clip and bend it into the shape shown below. Push the straight end of the clip through the cardboard and tape the clip to the cardboard.

2. Cut two strips of aluminum foil, each measuring one centimeter by four centimeters. Carefully smooth the foil. Push both pieces of foil through the bent end of the clip.

3. Place the cardboard over the glass container: This is your electroscope. The two pieces of aluminum foil will push apart from each other when there is an electrostatic charge on a nearby object.

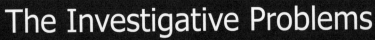

The Investigative Problems

How can I make an electroscope?
What does an electroscope detect?

Gather These Things:
- ✔ Rubber balloon
- ✔ Wool cloth
- ✔ Silk cloth or other fabrics
- ✔ Clear glass container
- ✔ Lightweight aluminum foil
- ✔ Assortment of rubber, plastic, and glass objects
- ✔ Cardboard square (10 cm. x 10 cm.)
- ✔ Medium-size paper clip
- ✔ Scissors
- ✔ Tape

To see how it works, blow up the balloon and use a piece of wool to rub the balloon. This will cause the balloon to acquire an electric charge. Slowly move the balloon until it touches the top of the wire. Watch the little pieces of foil. What happens to the pieces of foil when a charged balloon touches the wire on the top of the electroscope?

Discharge the electroscope by touching the wire on top of the electroscope. This will get rid of the extra electrons. Charge the balloon again and move it near the top of the wire, but don't let them touch. What happens to the piece of foil when a charged balloon is brought near the wire?

Discharge the electroscope again by touching the wire. Test the following objects: a wooden pencil, a green leaf, and a glass cup. Touch each test object with your hand, but don't rub them with anything. Bring each object close to the wire on top. Record what the two aluminum foil pieces do.

Discharge the electroscope by touching the wire after each test. Try charging some objects by rubbing them with another material. Things like rubber, paper, and plastic objects tend to hold a charge. Wool, nylon, and silk are likely to produce a charge on these objects. Try charging a variety of objects by rubbing them with wool, nylon, silk, or something else. Bring each object near the wire. Keep up with how you charged each object you test. Record whether or not the charge was detected by the electroscope each time.

Object tested	Rubbed with	Ch. Detected by Elec.
1		
2		
3		
4		
5		

The Science Stuff

You need to keep the foil in the container because it is so thin that the slightest breeze will cause it to move. You want the movement of the foil you observe to be only from the charge it receives.

Glass and cardboard not only shield from wind, but they are also good **insulators** of electricity. Things like glass, cardboard, rubber, paper, and plastic are good **insulators** of electricity, because electrons cannot move easily though these materials.

Electrons can move easily through steel, aluminum, and other metals. These materials are good **conductors** of electricity.

The electroscope is designed in such a way that electrons can only move through the paper clip and the aluminum foil (conductors), and not through the glass or the cardboard (insulators).

Recall that there is a magnetic field around a magnet, and there is also an **electric field** around a charged object (+ or -).

If a charged object that has a negative charge is brought near the wire, its electric field will push some of the loosely held (negative) electrons down the wire. Extra electrons will go into both pieces of the foil, making both pieces negative. As you know, the like charges will repel, so the pieces of aluminum foil will move away from each other.

If a charged object that has a positive charge is brought near the wire, its electric field will attract some of the loosely held (negative) electrons and they will move up the wire and out of the foil. This will leave the foil with fewer electrons and an overall positive charge. Two positive charges will repel and the pieces of aluminum foil will move apart.

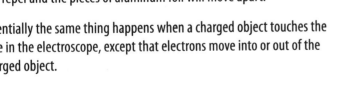

Essentially the same thing happens when a charged object touches the wire in the electroscope, except that electrons move into or out of the charged object.

An item has a positive or negative charge on it if the pieces of aluminum foil move apart when the item is brought near the electroscope. (You can't tell whether the charge is positive or negative.) The pieces of aluminum foil will not move if you bring an uncharged object near the electroscope.

Making Connections

This should remind you of how magnets and objects with a static electric charge are alike. Both have invisible fields around them. The **magnetic field** around a magnet can affect other magnets without touching them. Charged objects have an **electric field** around them that can affect other charged objects without touching them. Electrons (negative charges) are repelled by negative charges and attracted by positive charges, very much like two north poles (or two south poles) of magnets repel each other, and a north pole and a south pole attract each other.

Dig Deeper

Electroscopes have been used by scientists since the 1700s. Do some research on the history of electroscopes. This was an important piece of equipment for some scientists. How did some of the early scientists use them?

What Did You Learn?

1. Explain what might cause some of the electrons to get pushed down to the foil in the electroscope and make the foil move apart.

2. Explain what might cause some of the electrons to get pulled up from the foil in the electroscope and make the foil move apart.

3. Can an electron be pushed or pulled by the electric field around a charged object even though the object is not touching the electron?

4. What do we call materials that electrons can move through easily? Give two examples.

5. What do we call materials that electrons cannot move through easily? Give two examples.

6. How do you discharge an electroscope?

7. What are some ways in which magnets and objects with an electric charge around them are alike?

How Two Simple Ideas Became Huge

Static electricity and magnetism had long been suspected as being connected in some way. They both had invisible fields around them. They both obeyed the rule of "likes repel and unlikes attract." Until 1820, no one had found a way in which electricity and magnetism were connected. The discovery that there is a connection between electricity and magnetism turned out to be one of the most important scientific discoveries in history.

The first connection was made in 1820 while a science lesson was being taught. The teacher was Hans Christian Oersted, who was a professor at a university in Denmark. He had arranged to do some science demonstrations in his home for a group of friends and students. He built a battery from chemicals and sheets of metal to demonstrate how an electric current could heat a wire. He also brought a compass needle mounted on a wooden stand as part of some other demonstrations about magnetism.

During the demonstrations, Oersted noticed something unexpected. Every time the electric current was switched on, the compass needle moved. After the demonstrations were over, he and some of the students continued to play around with the battery and the compass.

Each time they made a complete circuit by connecting the battery to the wires, the compass needle would move. When the wires were disconnected, the compass needle would move back and point north again. There was something invisible affecting the compass needle when the current was on. The wire was acting like some kind of strange magnet.

Professor Oersted published the results of this experiment. In just a few years, other scientists, like Joseph Henry, had made powerful electromagnets by wrapping wires around a soft iron core and connecting the wires to a battery. One of the advantages of electromagnets was that they could be turned on and off. Another advantage was that they could be made much stronger than ordinary magnets. Henry built an electromagnet that could pick up over two thousand pounds.

Electromagnets led to the invention of the telegraph. Joseph Henry was the scientist most responsible for the long-distance telegraph. However, Samuel Morse was the person who got the patent for it. Morse also devised a code known as the Morse Code that was used to send messages.

Oersted had shown that a moving electric current could produce a magnetic field. Michael Faraday from England and Joseph Henry from America were two scientists who wondered if a magnetic field could make an electric current. In fact, they made this discovery at about the same time, independent of each other, in 1831.

Subsequently, it was discovered that there was a very simple way to make an electric current just using wires and magnets. Simply take the two ends of a piece of wire and twist them together, making a circular loop of wire. Hold two bar magnets apart with opposite poles facing each other so that magnetic lines of force form between them. Have someone move the connected loop of wire back and forth between the magnets. Every time the wire moves down and cuts through the magnetic lines of force, electrons in the wire are pushed in one direction. Every time the wire comes back up through the lines of force, the electrons are pushed in the opposite direction. (Even though you won't see or feel anything, electrons are moving back and forth in the wire.)

You just produced an A.C. electric current, where the electrons move back and forth. Obviously the current is very weak and you would get tired trying to keep the wire moving back and forth.

However, wind power, water power, and steam can be used to keep many loops of wires inside electric generators moving across magnetic lines of force. The loops in a generator actually rotate instead of moving up and down. Most of our A.C. electricity is produced by electric generators.

So, by 1831, two important scientific ideas had been discovered: (1) An electric current can produce magnetism, and (2) magnetism can produce an electric current. Years passed while research and technology based on these two little ideas continued. The work was done by both scientists and inventors. Not only were telegraphs invented, but by the late 1800s, electric generators had been built, along with light bulbs and other electric devices. Electric motors were also being built.

Our demand for electricity seems to constantly increase. Recall that electric energy cannot be created. It can only change from another form of energy. As the need for electricity continues to grow, we will need to look for more sources of energy. The most promising sources are the renewable ones, such as solar energy, geothermal energy, wind energy, and biomass. At the same time, we need to find better ways to conserve our nonrenewable energy sources, such as coal, oil, natural gas, and nuclear energy.

Discussion:

Make a list of things you use frequently that require current electricity.

What is an electric generator? Why does it have to turn in order to produce electricity?

The discoveries that (1) an electric current can produce magnetism and (2) magnetism can produce an electric current are the basis for much of our technology today. Who were the scientists responsible for each of the discoveries? List some of the technology we have today because of these discoveries.

Think about This

You complete an electrical circuit when you flip a light switch on. An AC household circuit would be too dangerous to experiment with, but a DC circuit using a small battery is a perfectly safe way to learn the basic concepts about series and parallel circuits. Do you think the appliances and lights in your house are wired in series or in parallel circuits?

Procedure & Observations

Connect your circuit as in the diagram. ➔

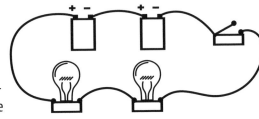

If your lights don't come on, check your wiring again. Be sure the insulation on the wire is removed at all the connection points. The current cannot flow through the insulation. Make certain that the exposed part of the wires is properly connected and the bulb is screwed into the socket. Be sure the switch is closed (in an on position). Test your light bulbs to make sure they are all good. You may also need to test your batteries to make sure they are all good.

1. When you get the lights to come on, examine the switch. A switch allows you to turn the flow of current off and on. Open the switch (open means to turn off). What happened?

2. Close the switch. What happened?

3. While the lights are all on, disconnect one of the wires to the bulb. What happened?

4. Reconnect the wire and then disconnect one of the wires to the battery. What happened?

5. Reconnect the wire. While the lights are all on, unscrew one of the bulbs. What happened?

6. Take one of the bulbs and sockets out of the circuit and rewire the circuit. Is the one remaining bulb brighter now?

The Investigative Problems

What is a series circuit?

What are the essential parts of an electric circuit?

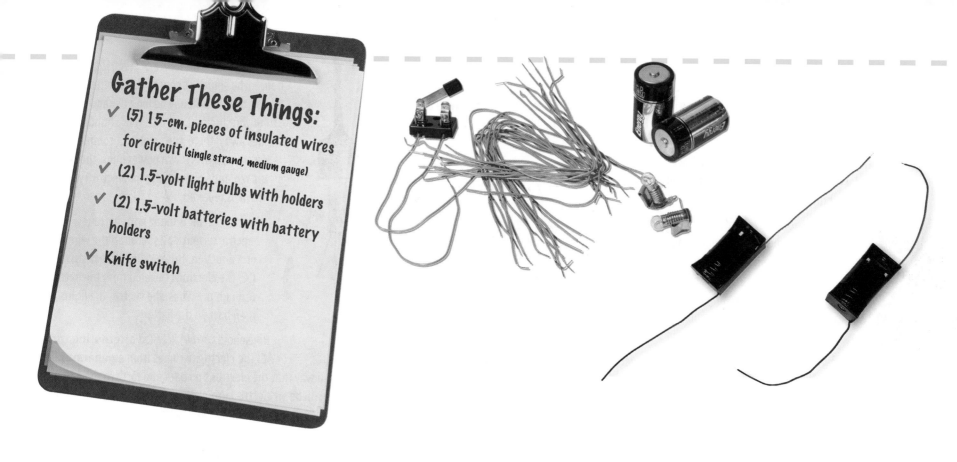

Gather These Things:

✓ (5) 15-cm. pieces of insulated wires for circuit (single strand, medium gauge)

✓ (2) 1.5-volt light bulbs with holders

✓ (2) 1.5-volt batteries with battery holders

✓ Knife switch

Using the following symbols, label the circuit diagram (or schematic) for a series circuit that you constructed.

Symbol	Label
	one cell
	two cells connected by a wire
	two-cell battery
light bulb	light bulb
switch	switch

Symbol	Label
or	one connecting wire
∿	a resistor
Ⓐ	an ammeter
Ⓥ	a voltmeter

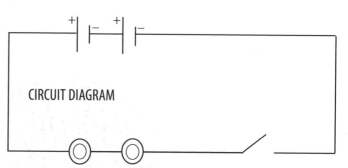

CIRCUIT DIAGRAM

Notice that a straight line represents one wire, but a line drawn with a right angle or two right angles is still one wire. The boxy shape keeps the diagram simple. Draw a circle around the source of the energy in the diagram. Draw a triangle around the things that use the energy.

The Science Stuff

Recall that **static electricity** occurs when electrical charges are produced by rubbing two substances together. Electrons don't move where there is static electricity unless there is a sudden discharge of electrons. This investigation used a different kind of electricity, known as **current electricity**, where the **electrons** continually move through a wire or other **conductor**.

In order for a current of electricity to flow, there must be a complete pathway or circuit. All electrical circuits need three things: a power source (like a battery), something to use the electricity (like a light bulb), and connecting wires. Most circuits will also have a 4th part, a switch, which allows you to turn off the current when it is not being used.

In this activity, the power source was the battery. The light bulbs used the energy. Wires connected the source to the bulbs. This circuit also contained a switch. Even if you have all the needed equipment, everything must still be connected so that you have a complete pathway through which the electrons can travel without a break.

The circuit you made is called a **series circuit** because there is only one pathway for the electrons to travel through. The circuit is incomplete if a light bulb is removed, a bulb is "shot," the switch is off, a wire is disconnected, or the battery is dead. If the circuit is incomplete, the current will not flow and none of the lights will come on. In a series circuit, there is no alternative pathway.

The electrons that move through the circuit get their energy from the battery. This energy was transferred from the battery to the light bulb by the electrons. Some of the electrical energy from the battery changed into light and heat energy in the light bulb. Heat energy is produced even though it is not needed and costs money.

Batteries are not the source of electricity to your house. The source of household electricity is generators, which may be long distances away from your house. Even if they are hundreds of miles away, there must still be a complete circuit. If a storm blows down power lines, the circuit is broken and you will have no electricity until the break is repaired.

When a circuit is made with batteries, the electric current flows through the wires in only one direction. This is called **direct current** or **DC**. The electrons traveled from the battery through the wires, the switch, the lights and then back to the battery.

Household current is called **alternating current** or **AC**. The electricity comes from a **generator** in such a way that the electrons go back and forth in the wires. There will be an electric current regardless of whether the electrons flow in only one direction or whether they flow back and forth.

Making Connections

Ordinary batteries (DC current) use up their stored energy after several hours. Rechargeable batteries are more expensive than ordinary batteries, but when they go "dead," they can be plugged into an electric charger and used again.

There has been a renewed interest in battery-operated cars and in hybrid cars that use both batteries and gasoline. The batteries would have to be recharged when they got weak. Recharging weak batteries would have to be more frequent than refueling with gasoline.

Dig Deeper

Use electric symbols to draw the story of Electron's journey: Electricity was pushed out of a 3-cell battery (12 volts) with lots of energy. He traveled easily through a copper wire. Then he entered a lamp and passed through the lamp's filament. It took a lot of effort to squeeze through the filament, and it cost him a lot of his energy. Then he was back in a copper wire again where his journey was easy for a little ways. Before he knew it, he was back in another lamp, pushing through the filament and losing most of his energy. Suddenly he came to a complete stop, along with everyone else. The switch was open. As soon as the switch was closed, Elec continued his journey back to the battery. He was full of energy again by the time he passed through the last cell. Write your own story and see if a classmate can draw the electron's journey.

Do some research on the advantages and disadvantages of an electric car. Write a summary of your findings.

What Did You Learn?

1. What kind of circuit is in your house — series or parallel?

2. Were two bulbs wired in series just as bright as one bulb?

3. What is the purpose of a switch?

4. Why did removing one bulb stop the other light from coming on?

5. What four things are found in most complete circuits?

6. Name a few reasons why the lights might not come on after you connected everything.

7. Draw a circuit diagram of the circuit you made. Label the parts and count the wires.

8. What kind of current is produced by a battery?

9. What kind of current is produced by a generator and used by most household appliances?

10. What unwanted form of energy was produced in the light bulb?

11. The chemical energy in the battery changed into electrical energy. What did the electrical energy change into in the light bulb?

Let there be light!

Is a Parallel Circuit Better Than a Series Circuit?

Think about This
You have learned that in a series circuit, when one light goes off, they all go off. Do you think there is a better way to wire the lights and appliances in your house? How do you think this could be done?

Michael Faraday (1791–1867) was an English chemist and physicist who contributed to the fields of electromagnetism and electrochemistry.

The Investigative Problems
Why can you turn off one light in your house while the other lights stay on?
What would be some problems with wiring everything in your house to a series circuit?

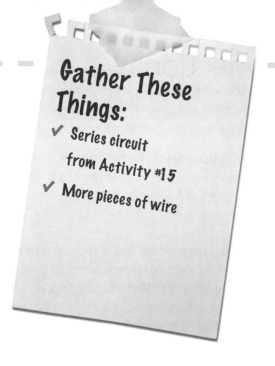

Gather These Things:

✓ Series circuit from Activity #15
✓ More pieces of wire

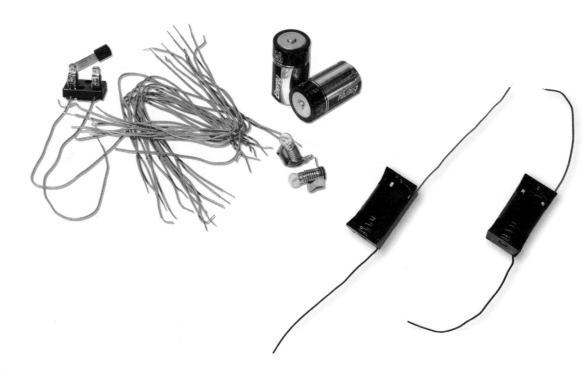

Procedure & Observations

Start with the same circuit that you used in the previous lesson. Check your circuit to be certain that all bulbs are working.

Remove wires from the light sockets and reattach the wires as in the diagram.

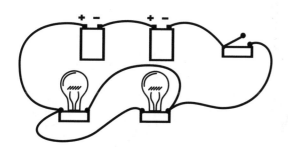

Leave the batteries connected like they were. The batteries are wired in a series connection, but the lights are wired in parallel.

When all the lights are on, unscrew one bulb. What happened? Is the remaining bulb brighter than before?

Screw it back and unscrew the other bulb. What happened? Is the remaining bulb brighter than before?

Unscrew both bulbs. What happened?

Explain how this parallel circuit is wired differently than the series circuit.

Make a circuit diagram of the circuit used in this activity.

The Science Stuff

Batteries, wires, light bulbs, and switches can be connected in either a parallel circuit or a series circuit. There are advantages and disadvantages of each kind of circuit.

In a parallel circuit, there is more than one pathway for the electric current to travel back to the battery. Therefore, when one bulb is removed, the current can still find another pathway back to the battery. In a series circuit, if one light bulb or appliance goes off, everything in the circuit will go off.

Multiple lights in a parallel circuit will burn as bright as one bulb, because the current does not have to travel through all the light bulbs. In a series circuit, the current has to travel through every light. If several lights are on, they will all be very dim.

Each light in a parallel circuit gets about the same amount of current that the battery gives off. The downside is that the batteries won't last as long in a parallel circuit as in a series circuit.

Most houses have parallel circuits for their lights and other electrical devices. For most household electrical devices, it is more convenient and useful to use parallel wiring, so when one light goes off, the others will stay on.

We used batteries (DC current) to provide the electricity for the past two investigations, because household electricity would be too dangerous. Batteries do not provide the electricity for electric lights and appliances in your house. This electricity comes from an **electric generator** (AC current).

Recall that an electric current can be produced whenever a connected wire cuts through magnetic lines of force. That is what an electric generator does, except the wire is looped many times around a moving part known as an armature. As the armature turns halfway around, electrons are pushed through the wire in one direction. When it makes another half turn, the electrons are pushed in another direction. This changing back and forth makes an AC current.

Making Connections

Sometimes a light is connected in series to an electrical device. The light will be on as long as the device is working. The light will go off if the device stops working. This arrangement lets people know that if the light is off, the device is probably not working.

Strings of decorative Christmas tree lights used to be wired in series, because they didn't use much electricity. Most of these lights are now wired in parallel, because if one bulb burns out, all the bulbs will go off. It is a lot of trouble to find the bad bulb.

Dig Deeper

Locate the schematic diagram on a small radio or other electrical device. A schematic shows all the electrical connections and pieces of equipment in the radio by means of symbols. Find a good book on electronics and see if you can identify the symbols on the schematic. Be sure to leave the schematic on the radio. If the radio ever needs to be repaired, the electrician will need the schematic to know how to reconnect the parts.

What is a generator? Find out where the generators are that supply electricity to your home. What causes these generators to turn and produce electricity? Locate these generators on a map. Locate your house on the map, and determine how far away the generators are from your house. What is needed to keep the electric current moving through the wires for long distances?

After Michael Faraday found that electricity could be produced from magnetism, he constructed a device that he called an electric dynamo. Dynamos were an early version of the modern power generators. Read a book about this amazing scientist or do some research about his life. Summarize what you have learned.

Hospitals usually have a set of generators as a back-up source of electricity in case something causes the electricity to go off. Some families keep a generator near their homes in case of a prolonged power outage. However, generators have to have their own source of energy. It wouldn't help to have a generator unless there was a source of energy to make it work. What are sources of energy that could make a generator work? Also, one working generator would probably only have enough power to operate a few electrical devices. Make a list of the things in your house that operate on current electricity. What would you consider the three most important things to connect to the generator if your electricity was off for a month?

What Did You Learn?

1. What would happen to the brightness of the lights if you added more lights to the same parallel circuit?

2. Which was brighter — two lamps wired in series or two lamps wired in parallel?

Refer to the following circuit diagram to answer questions 3–6.

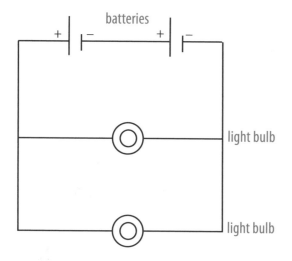

3. Are the lights wired in a series or a parallel circuit?

4. How many lights are shown in this circuit?

5. How many dry cells are shown in this circuit?

6. How many wires are used in this circuit?

7. What are two reasons why household circuits are wired in parallel?

8. What is one disadvantage of connecting several lights to a battery in a parallel circuit?

9. As an electric generator produces electricity, connected loops of wire cut through what?

Howdy, Partner!

Writing Project: A Summer Day in the 1800s

You were probably born in a time when everyone you knew used household electricity. Imagine you had been born between 1800 and 1900 on a farm in America. On a typical summer day, there would be no school, no electricity, no running water, and no time to be bored. A lot of good books have been written around this setting.

Read a book based on this time frame, and then write a description of how you might have spent a summer day. Give the name of the book you read. Who was the author?

71

INVESTIGATION #17

Think about This

We have learned that all electric circuits must have a source of energy (batteries/generators), something that uses the energy (light bulbs, appliances, etc.), conductors (usually wires) to connect them, and usually a switch.

We would like an easy way to understand how the electric current is related to the voltage (from the battery/generator) and the resistance (in the light bulbs, appliances, etc.).

The Investigative Problems
How are current, voltage, and resistance in a circuit related?
How does a change in one of these affect the others?

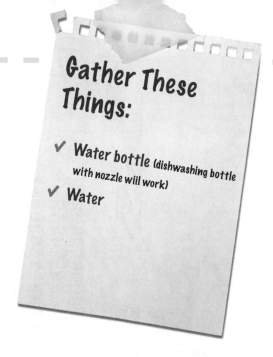

Gather These Things:

✔ Water bottle (dishwashing bottle with nozzle will work)

✔ Water

Procedure & Observations

Fill the bottle with water. Go outside for the remainder of this activity.

Open the nozzle all the way and squeeze on the bottle as hard as you can. Observe and record what happens. Answer the questions as you do each part of this activity.

Close the nozzle so that it is almost completely closed, but not quite. Squeeze. Observe. Record what is different about the first and second squeeze.

Leave the nozzle where it is and try to make the water come out as fast as it did the first time. Describe how hard you had to push to get the water to come out faster on the third squeeze.

The water flow can be compared to the current in an electric circuit. The push can be compared to the voltage. The nozzle setting is like the resistance. We will try to make some comparisons to help you better understand the three basic terms related to an electric circuit — voltage, current, and resistance.

What happens to the current (water flow) when you decrease the voltage (how hard you have to push)? What happens to the current (water flow) when you increase the voltage (how hard you have to push)?

When you change the resistance (nozzle setting), how does that affect the voltage (how hard you have to push)?

What will happen to the current (water flow) if you keep the voltage (how hard you have to push) the same and increase the resistance (close the nozzle partway)?

Try to fill in the blanks in the chart below. Think about the bottle of water as you do this.

Relationships of current, voltage, and resistance		
Volts (How hard to push)	Resistance (Nozzle setting)	Current (Water flow)
stays same	increases (closes)	
stays same	decreases (opens)	
increases		stays same
	stays same	increases
	stays same	decreases

The Science Stuff

Voltage, resistance, and current are three basic units of current electricity. All three are related to each other within a circuit. If one of these changes, it will affect one of the others.

A high current is very dangerous, because the damaging effect of an electric shock comes from high current. This means the circuits with the lowest resistance would have the largest current and would be the most dangerous. Recall that a bigger water flow came out of your bottle when the nozzle was opened up all the way. This can be compared to a large current in an electric circuit where there isn't much resistance.

A resistance is something that doesn't let electricity flow as easily as a good conductor would. The filament in a light bulb is a resistor. It is hard for the electricity to flow through the filament. Resistance is what causes the filament to give off heat and light. The nozzle represents the resistance in the circuit. The electrons moving through the circuit meet resistance in other devices, such as refrigerators, radios, and electric heaters.

The source of the electricity is usually a battery or a generator. These devices determine the amount of voltage in a circuit. The push you exerted on the bottle can be compared to the voltage.

There is a mathematical formula that shows how volts, current, and resistance are related. The formula can be written as volts = current (amperes) x resistance (ohms). It can also be written as current = volts/resistance.

Suppose in a DC circuit the current is 5 amps and the resistance is 2 ohms. Using the first formula, we find the battery is sending out 10 volts.

When the resistance stays the same, an increase in the voltage (push) also increases the current (water flow). If the resistance (close nozzle partway) increases and the voltage (push) stays the same, the current (water flow) will be smaller. If the resistance is low (nozzle opened up) and the voltage remains the same, the current (water flow) will be greater.

Making Connections

In your house, sometimes the circuit becomes overloaded, such as when there are too many devices using electricity at the same time or when there is a short circuit. A circuit breaker or a fuse will usually shut down the circuit. Otherwise, the wires might get too hot and start a fire. Circuit breakers and fuses are important safety features in houses.

Circuit boxes are dangerous and should be handled by adults only.

Dig Deeper

Make a list of safety rules for using electricity in your home and for being around downed power lines after a storm.

A DC voltmeter can be used to measure the amount of voltage across one of the light bulbs in your circuit. Use one of the series or parallel circuits you assembled before and make sure all the lights are on. Without disconnecting any of the wires, add wires from a voltmeter to the two sides of a light socket.

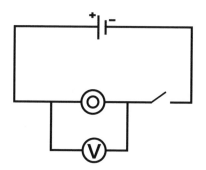

The voltmeter will measure the difference in the voltage going into the light and the voltage leaving the light. The more voltage the light uses, the greater the reading on the voltmeter will be. You can also connect the voltmeter across a dry cell in the same way. The voltmeter will show the difference in the voltage entering the dry cell and the voltage leaving the dry cell. Determine the voltage of each light bulb and each dry cell in your circuit.

All three units can be measured by using the correct meter or by doing some math. A voltmeter measures the voltage. An ammeter measures the currents in amperes. An ammeter is wired into the circuit in series. Reconnect the series circuit you did earlier. Connect the voltmeter across the battery and connect the ammeter in the circuit in series. Then calculate the resistance in ohms by dividing the volts by the amperes. (This is not as hard as it sounds. You just have to have the right kind of meters.)

What Did You Learn?

(1–4) Analyze the analogy of the bottle to an electric circuit.

1. What does the "push" you made represent?

2. What does the nozzle represent?

3. What does the water coming out of the nozzle represent?

4. Explain this analogy in your own words.

Narrative: Nonrenewable Conditions for Nonrenewable Fuels

Joey and Joanne sat down by their parents, Dr. and Mrs. Houston watching a news commentator broadcasting from the proposed site of a new wind-powered electric station. He went on to talk about the advantages of wind power. Wind does not pollute the air and the power is renewable.

"What does renewable mean?" Joey asked.

"It means wind is a source of energy that will not be used up for a very long time," his mother answered. "Some nonrenewable energy sources may only last for a few more hundred years. "

"What are our nonrenewable energy sources?" Joanne asked.

"These are mostly the coal, oil, and gas reserves found around the earth," Dr. Houston answered. "Once they are used up, there may not be any more of these resources."

"It looks like more fuels could just be made inside the earth," Joey said.

"No, there had to be some unusual conditions in order for these fuels to form. The main thing was that huge numbers of plants and animals had to be trapped and buried before they had a chance to rot and decay," Dr. Houston said.

"So, how did fuels form?" Joanne was very interested in knowing more.

"This is what I believe happened. Thousands of years ago, there was an abundance of plants and animals on the earth. The earth's average temperature was probably more than it is today. I believe even the polar regions were warmer at that time than they are today. Then three things happened all at the same time.

"Large drops of rain began to fall for the first time in history. There had been misty type showers before, but nothing like the downpour that was falling. Then water began gushing out of subterranean cracks in the ocean floor. The disruptions in the underground rocks caused volcanic eruptions and earthquakes all over the world. As these things happened, trillions of plants, land animals, and ocean life were trapped and buried by such

things as mudflows or volcanic debris. These catastrophes may have continued for hundreds of years."

Joey ran to his room and came back with a large, colorful book he had checked out of the library. It was about one of his favorite subjects — fossils. There was a series of pictures in the book that was different from his dad's explanations.

"Look at these pictures, Dad. See the fish swimming in the ocean. In the next picture, the fish dies and sinks down into the ocean mud. In the last picture, millions of years pass and the fish turns into a fossil. Is this how some fossils formed?"

"Probably not," Dr. Houston replied. "As soon as a fish dies, it either rots and decays or some other animal comes along and eats it. There are trillions of animals and plants dying today, and it is doubtful that many of them are in the process of becoming fossils. The only living things that might turn into fossils are ones that get trapped in a mudflow, volcanic debris, lava flow, or something else where they would get buried quickly and be cut off from oxygen."

"We know there were thousands of buffalo killed in the American West. They either decayed or were eaten. Some of the buffalo bones were left on the ground for several years. Even these bones have turned to dust by now. But no buffalo fossils formed unless the animals or the bones were quickly buried by thick layers of earth under just right conditions."

"But, Dad," Joanne said, "is it possible that fish could die and fall into the ocean sediment before they decayed or got eaten?"

"I suppose there are conditions under which that might happen, but it's hard to explain the huge amount of fossils on earth in terms of a rare event. Let me give you something to think about. Have you ever heard of nonrenewable fossil fuels?"

"Sure, I did a report on it."

"So, I'm sure you remember that fossils fuels are things like coal, oil, and gas," Dr. Houston said.

"Yes. I learned that coal is mostly the remains of plants, and oil and gas are the remains of both plants and animals."

"I'm glad you were paying attention. Most scientists agree on this much, but there are two main explanations about how the fossil fuels formed. One group thinks they formed gradually over millions and millions of years. Another group thinks they formed much more quickly than that. They believe most fossil fuels formed as a result of a worldwide series of tremendous cataclysmic events that will never happen again."

"Like Noah's Flood!" Joey said.

"Exactly," Mrs. Houston agreed as she finished the explanation. "Look at this picture in your book of a fossil fish. All the bones were preserved and even show details.

"Which explanation sounds more reasonable? One, the fish was buried quickly and deeply by some kind of sudden upheaval, preventing decay and keeping other animals from eating it, or two, the fish died, sank into the sediment of the ocean, and stayed there for a long time without decay and without little creatures in the mud affecting the bones until it became a fossil."

"It looks to me like the first explanation is the best one," Joey said.

"I think so, too," Joanne agreed.

Discussion:

Which of the two explanations for how the fish became a fossil do you think is most logical? Why do you think so?

Conserving and saving our nonrenewable fuels is becoming more and more important. What are some of the ways we can all conserve energy? Write about one of these ways.

What are some of the alternative energy sources scientists and engineers are .....investigating? Write about one of these sources.

Pause and Think: Future Sources of Electricity

There are many sources of energy that can turn a generator and make electricity. Fossils fuels are one of the sources of energy for generating electricity. Some other ways that have been proposed to generate electricity include using energy from wind, water, geothermal sources, biomass, tidal power, and nuclear power.

Solar batteries, chemical batteries, and hydrogen cells are other ways of producing electricity.

We will look at a few of the ways in which electricity can be produced, although there are many more than the few we will examine.

One of the big problems facing your generation is how to conserve our nonrenewable fossil fuels, such as coal, oil, and natural gas. We also need to find alternative sources of energy from the sun and the earth. At the same time, we must balance our energy demands with standards for clean air and water. We especially need informed, creative men and women to be involved in finding ways to meet the energy needs of the world. Is this something you might want to prepare to do in the future?

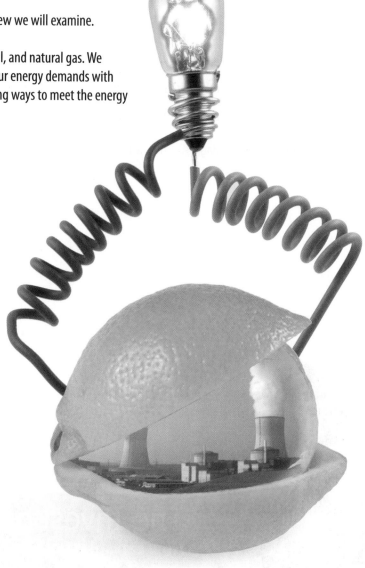

Society's need for electricity seems to just keep on growing, in spite of huge hydroelectric dams and other energy sources. Nuclear energy has the potential to be a valuable source of energy, but if not properly managed, it could become a dangerous or environmental problem. The United States has plenty of coal to produce electricity for many years to come, but many people fear more coal-powered generators would cause air and water pollution. It's going to take a lot of smart, creative ideas to balance our need for electricity with our needs to preserve and protect the environment.

Think about This

Lori's dad was putting lights on both sides of the driveway.

"Where are you going to plug in all these lights?" Lori asked.

"I don't need a plug," her dad said. "These are solar lights. The batteries are charged by the sun during the day, and the charged batteries keep the lights on all night."

"But what if the sun gets covered up with clouds one day?" Lori asked.

"There may be times when they don't work, but they will work most of the time. At least we don't have to pay a solar bill to use the sunshine."

Is solar energy something we could use more often?

The Investigative Problems
How can solar energy be changed into electric energy?
How can solar energy be changed into heat energy?

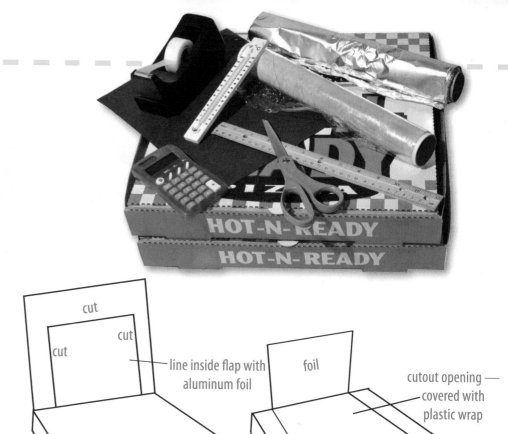

Gather These Things:

- ✓ Calculator with solar batteries
- ✓ Two medium-size pizza boxes
- ✓ Aluminum foil
- ✓ Thermometer
- ✓ Plastic wrap
- ✓ Metric ruler
- ✓ Black construction paper
- ✓ Tape
- ✓ Scissors

cut

cut

cut

line inside flap with aluminum foil

foil

cutout opening — covered with plastic wrap

① ②

Procedure & Observations

Look at a solar-powered calculator. It should have the word "solar" written on it. See if you can tell where the solar batteries are located and how they can be recharged. Turn the solar-powered calculator on and enter several numbers. Cover the solar batteries with a soft material that will keep the light out. Wait a few minutes and observe the screen.

Some solar-powered calculators will quickly stop working after the light is blocked. Other models will continue to work for much longer. What happens to the numbers on your screen? Remove the covering over the solar batteries. What happens?

Now we will look at one way to convert solar energy into heat energy by building a simple solar cooker. To make a solar cooker, close the lid of a medium pizza box. Use your ruler and measure six centimeter from the front and side edges of the lid. Draw three straight lines that are six centimeter from the front and six centimeter from each side. This will make a separate flap that can be lifted up at different angles.

Cover the inside of the flap with aluminum foil. Tape the foil to the flap. Gently rub the foil until it is shiny and smooth. Lift the foil-covered flap and cover the opening in the lid with

plastic wrap. Use tape to hold the plastic wrap in place. Cover the bottom of the box by putting down a few sheets of black construction paper. This will be Box 1.

Box 2 will be the control group. It will be used to compare with Box 1, which will be the experimental group.

Place a thermometer inside both pizza boxes. Tilt the aluminum-covered flap of Box 1 so the sun's rays are reflected into the box. Tilt the lid of Box 2 at the same angle and place both boxes side by side in bright sunlight.

After five minutes, read the thermometers. Record the temperature inside each box. Take three more readings, each one five minutes apart. Record these temperatures. Look carefully at the thermometer readings from each box. Subtract the first temperature from the last temperature in Box 1. Subtract the first temperature from the last temperature in Box 2. Which box got hotter?

The Science Stuff

Solar power refers to taking sunlight and changing it into another source of energy. Solar batteries can change light into electric energy. Sunlight can also be used as a source of heat energy.

The solar batteries in the calculator change light into electricity. Both sunlight and artificial light will cause solar batteries to work. The kind of solar lights Lori's father was setting up also contained solar batteries. The batteries in these lights are charged by the sun during the day. There is enough stored energy to keep the lights on during the night.

Most of our energy comes to us from the sun, either directly or indirectly. Some of this energy is visible light and some of it is invisible. Only a tiny fraction of the sun's light reaches the earth. However, the energy transferred from the sun to the earth is the cause of many different kinds of changes on the earth's surface.

One type of invisible wave from the sun is known as an **infrared wave**, which is sometimes called a heat wave. These waves are generally changed into heat when they strike something. Another invisible wave from the sun is an **ultraviolet wave** (or UV wave). Ultraviolet waves are not all the same length, but UV waves are shorter than visible light and infrared waves. They are longer than X-rays and gamma rays, which are also electromagnetic waves.

Most of the energy that comes to the earth from the sun is in the form of **electromagnetic waves**, which have many different wavelengths. These waves can travel through space, as well as through the air. The shorter the wavelength, the more energy the wave carries. The very shortest waves are also the most dangerous.

Making Connections

Ultraviolet waves (UV) have some interesting uses, but they can also cause serious harm. They have enough energy to break chemical bonds in the skin. Too much exposure to UV waves can cause your skin to become sunburned or could even cause skin cancers. Skin damage is usually found on places where the skin has been exposed to the sun for a long time.

Doctors recommend that you put sunscreen on your skin when you are exposed to sunlight to prevent damaging your skin. Tanning beds use UV waves and should be used with caution, if at all.

Green plants have an amazing ability to take energy from the sun and store it in food that is manufactured in their cells. Researchers have not been able to do all that green plants do, but they have learned a number of ways to harness and use the power of the sun.

Dig Deeper

Plan an investigation to test and compare your solar cooker with the control pizza box. Start by making two sandwiches as nearly alike as possible, by putting a large marshmallow between two graham crackers. Include a way to add a piece of chocolate. Describe the amount of melting of the marshmallow and the chocolate for each sandwich. Write about your investigation in such a way that someone who had not seen it would understand what you did. Include drawings if you think that would be helpful. If you can think of a way to improve the solar cooker, use your version in this investigation.

Infrared waves, or heat waves, are another kind of invisible wave that comes from the sun. Night vision goggles pick up infrared waves from things that give off heat and let people see in the dark. Both ultraviolet waves and infrared waves can be photographed with special equipment, even though they are invisible. Find out more about how these waves can be photographed, even on a dark night. How do police and the military use infrared equipment?

Demonstrate how UV waves cause fluorescent materials to glow and give off light. Purchase some fluorescent paint and a UV lamp (a "black light"). Paint a picture using both fluorescent paint and regular paint. Hang it on the wall of your room. Turn off the regular lights and turn on the UV lamp. Notice the differences in the painting when seen under both regular lighting and a UV lamp.

Use a reference book to find more information about energy that is carried by electromagnetic waves. Notice there are many different waves with different wavelengths. The ones with the shortest wavelengths are the most dangerous. Try to arrange the different electromagnetic waves from the longest wavelengths to the shortest wavelengths. Can these waves travel through both space and the atmosphere? What protects the earth from dangerous levels of the shortest waves?

What Did You Learn?

1. Name an invisible wave that comes to the earth from the sun, has a wavelength longer than red, and can be felt as heat when it is absorbed by an object.

2. When an object absorbs light, what energy change takes place?

3. Give at least two ways that solar power can be used.

4. How can you recharge a solar battery?

5. Does sunscreen lotion help to prevent ultraviolet waves from being absorbed by your skin?

6. What are some beneficial uses of ultraviolet and infrared light?

7. What kind of electromagnetic waves cause a sunburn, are known as "black light," and cause fluorescent paints to glow?

8. Which kind of electromagnetic wave is considered most dangerous — one with a very long wavelength or one with a very short wavelength?

9. Can electromagnetic waves travel through empty space?

10. When do infrared waves change into heat energy?

The man in the picture spent too much time being exposed to UV rays from the sun. Notice the areas where his glasses blocked the UV rays, such as across his nose, around his eyes, and from his eyes to his ears. He could have prevented UV rays from damaging his skin by applying some lotion that blocks UV rays.

Wind or Water Energy

Think about This

Greg and his family were spending their vacation camping near Lake Mead. His mom explained that the lake was formed when Hoover Dam was built across the Colorado River more than 70 years ago. Its waterpower has been a source of electricity to the surrounding area for many years. The dam was so huge, Greg was sure that it could provide electricity for everyone in the surrounding states. "Not quite," Mom said. "It serves a lot of people, but at times it still isn't enough, and steam-powered stations have to be used."

Do you know where your electricity comes from? It may be from a hydroelectric power dam, a steam-driven power plant, a nuclear power plant, or another source.

In places where there is lots of wind, electricity can be generated from wind power. Using wind power is not a new idea. It has been used for hundreds of years in some places as a way of providing power for grinding and lifting. How much weight do you think a homemade windmill can lift?

Hoover Dam, originally known as Boulder Dam, is a concrete arch-gravity dam in the Black Canyon of the Colorado River, on the border between Arizona and Nevada.

Procedure & Observations

1. Cut a 25-cm. by 25-cm. square from the cardboard. Be sure this material is flexible and sturdy enough to not tear easily. Tape the pattern over the square and cut through the pattern and the cardboard. Carefully mark the screw holes on the cardboard and push the end of a nail through each mark.

2. Use the nail to make an indentation in the center of the end of the dowel. (You can pre-drill a small opening for the screw if you have a thin drill.) Turn the screw about halfway into this indentation, and then unscrew it and take it out. If the wood starts to crack, wrap several layers of tape around it tightly.

3. Now push the screw into each screw hole on the cardboard. Hold the windmill and the screw and push the screw into the opening in the end of the dowel. Use a screwdriver to turn the screw until it is firmly in place.

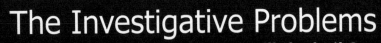

The Investigative Problems
How can windmills and water mills be built?
How much weight can a small windmill lift?

Gather These Things:

✔ 10-in. dowel — 5/16 in.

✔ Pinwheel pattern
(see Appendix in Student Journal)

✔ Very thin screw   ✔ Tape

✔ Hair dryer   ✔ Nail

✔ Screw driver   ✔ Heavy thread

✔ Small paper clip   ✔ Items to pick up

✔ Large paper clamp

✔ Glossy, lightweight cardboard or poster board

4. Make an indent somewhere near the center of the dowel where you can tie the thread on. Cut a piece of thread long enough to go from the floor to the table (plus a little extra). Tie tightly and secure with tape.

5. Slide the dowel through the open space of the paper clamp. Try to balance the clamp between the windmill and the weight. The clamp will give you a place to hold the windmill during the investigation.

6. Bend paper clip out to make a hook and tie onto the end of the thread. Use a hair dryer to blow the windmill. Weigh items that could be picked up, or rank them from lightest to heaviest by estimating.

Work with a partner, so that one person can hold the windmill and the other person can hold the hair dryer.

Put an item on the hook. Start small. Stand at least a meter apart and move closer together if you need to. Blow on each other's windmills with a hair dryer to see if it will lift the item. If your windmill doesn't lift the item, make sure the thread isn't slipping. Your goal will be to lift the item from the floor to the table.

Record the items you were able to lift. Tell if you could lift them completely or just part of the way. Mark if you could not lift it at all. Weigh each item or rank them from lightest to heaviest.

Item	Able to lift how high?	Not able to lift	Weight or rank
1			
2			
3			
4			
5			
6			

The Science Stuff

Recall that windmills and water mills make use of mechanical energy. The energy from moving wind and water has been used for centuries to grind corn and other food, to lift water, and to move other heavy objects.

For years, **hydroelectric dams** have used a type of water wheel, known as **turbines**, to turn electric generators and produce electricity. More recently, engineers have found ways to generate electricity from wind power. As the blades of the windmill turn, they can also turn electric generators.

Wind energy depends on wind, and sometimes it isn't windy. Electrical energy can be stored for times when the wind isn't blowing, or backup energy sources can be available.

The energy from sources such as wind, water, ocean tides, geothermal sources, and biomass are known as **renewable energy** sources. There is a great need for scientists and engineers to search for better ways to use these energy sources.

Most of our energy sources of fuels and electricity today are from **nonrenewable** sources, such as coal, oil, natural gas, and nuclear energy. The conditions under which these energy sources formed are no longer present. Therefore, we would not expect that they would continue to be produced in the earth by natural means. We must plan for alternative ways to produce energy, because the nonrenewable sources may run out someday.

Making Connections

Areas of the northeast United States have excellent places for wind farms, because they have strong winds most of the time. Texas, California, and places in the Midwest have also begun to build wind farms. However, wind resources have not been fully used in these areas. Much of the southeastern regions tend to have still, humid air and are not good candidates for wind farms.

It would not be practical to build wind farms in regions that don't have regular strong winds. Most ocean property does have regular winds, but landowners often object to windmills that spoil the beautiful scenery of the ocean. Finding solutions that please everyone will not be easy.

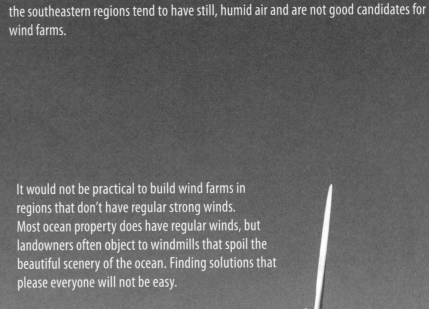

What Did You Learn?

1. What kind of energy is used by windmills?

2. Current electricity is produced when what kind of equipment is turned?

3. What can be done to provide electricity on a wind farm when the wind isn't blowing?

4. Why are wind farms not a good idea everywhere?

5. Name some renewable energy sources.

6. Name some nonrenewable energy sources.

Dig Deeper

Wind farms are a possible source of electricity for the future. Even though the wind is free, not everyone wants windmills in their area. Do some research and find the advantages and disadvantages of wind farms.

Do some history research about how Americans used wind power or waterpower. Windmills and water mills were an important part of our history.

What is biomass? What are some possible sources of biomass? There are certain kinds of micro-organisms that feed on certain kinds of biomass to produce something very similar to crude oil. See if you can find more information about this.

Think about This

It was December of 1938. Dr. Lise Meitner sat in her room in Sweden eagerly reading a letter she had received from her science partner, Dr. Otto Hahn, in Berlin, Germany, about an experiment in which uranium was being bombarded by neutrons.

Lise Meitner was part of the team that discovered nuclear fission, an achievement for which only her colleague Otto Hahn was awarded the Nobel Prize.

Her escape from Germany still seemed like a bad dream. Dr. Meitner was one of the top research scientists in the world, but she had been forced to flee for her life because Hitler's vision of building a "master race" included getting rid of Jews. Hitler was convinced that Jews and certain other races were hopelessly contaminated with inferior genes and had to be separated or eliminated from society. Lise had overcome prejudices many people had for women scientists, but she had no way to overcome Hitler's prejudice against her Jewish ancestry.

Lise was very happy when her physicist nephew, Otto Frisch, decided to visit her during the Christmas holidays. After catching up on family news, she immediately engaged Otto in a discussion about the unanswered questions of her nuclear research experiment.

The Investigative Problems
$E = mc^2$ **What do these symbols stand for and how do they impact science today?**

Dr. Hahn wrote that one of the substances produced during their experiment seemed to be barium. He didn't think this was possible because that would indicate the uranium nucleus was breaking apart. The idea of an atom splitting in half didn't fit with any of the current mainstream nuclear theories.

Going for a walk through a fresh layer of snow, the two scientists let their imagination freely examine new possibilities.

Sitting on a snow-covered log, she pulled complex formulas and measurements from her memory. Lise and Otto did their calculations, using scraps of paper and a pencil found in Lise's coat. They first calculated the energy that would be released if a uranium atom broke into two pieces, and found that each atom would release the incredible amount of approximately 200 million electron volts during the process. This seemed unreasonable, since the burning of a typical fuel only releases about 10 electron volts per carbon atom.

Lise began to think about Einstein's theories that matter and energy were related. She knew he believed it was possible for a small piece of matter to change into a large amount of energy. If Einstein was right, it would be possible to account for these massive amounts of energy. This was exactly what his famous equation $E = mc^2$ predicted many years ago. However, it would mean that a small amount of the uranium atom had actually disappeared and changed into a great deal of energy.

More calculations were done to determine that the mass of the original uranium atom was slightly more than the mass of the new products. The final calculation was to apply Einstein's formula and determine the amount of energy that would be released from the small amount of missing uranium. The stunned scientists could only stare at their answer: roughly 200 million electron volts, the same amount of energy that would be released if a uranium atom split apart!

Lise and Otto thought about the importance of what they had discovered. Their explanation was that uranium atoms were splitting in half as neutrons bombarded them. In the process, small amounts of matter were disappearing and being converted into huge amounts of energy. For now, they were the only two people in the whole world who had figured out how nuclear energy was being released from uranium atoms. They had no idea how this simple theory would change the world forever.

Through a very unusual set of circumstances, the news of their discovery was brought to a science convention in America by the famous scientist Neils Bohr. One of the scientists who was at the convention was Enrico Fermi from Italy. He had received a Nobel Prize for his work on uranium. He and his Jewish wife, Laura, and their children left directly from the award ceremony to move to America. Fermi became head of the first phase of the project that built a self-sustaining nuclear reactor.

In July of 1945 the first nuclear bomb was tested and exploded in a desert in New Mexico. Later that year, two nuclear bombs were dropped on the two main military bases in Japan, finally ending World War II.

Atomic bombing of Nagasaki, Japan, on August 9, 1945. The image was taken from one of the B-29 Superfortresses used in the attack.

Dig Deeper
Choose any character in this story and do some more research about him or her. Think of a creative way to tell about his or her life and what he or she did.

What are some of the peacetime uses of nuclear energy today?

What Did You Learn?

1. What did Lise and her nephew think they had figured out about what was happening when uranium was bombarded by neutrons?

2. Were Lise and her nephew looking for a way to build a new kind of weapon from uranium?

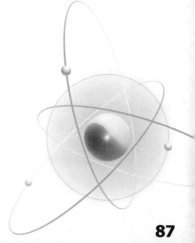

Calorie — A unit for measuring heat.

Conductor — A material that lets electrons move easily through it.

Concave lens — A curved, transparent lens that is thinnest in the middle and thicker around the edges.

Convex lens — A curved, transparent lens that is thickest in the middle and thinner around the edges.

Electric circuit — A complete pathway for electrons to travel where they can start at one place and get back to the starting place.

Electric current — A stream of moving electrons that transports energy from one place to another.

Electric resistance — The property of a substance that causes it to resist the flow of an electric current through it.

Electromagnet — A coil of wire wrapped around an iron core and connected to an electric current.

Electrons — Tiny particles with a negative charge that move around the nucleus of an atom.

Electroscope — An instrument that can detect an electric field.

Energy — Anything that can change the condition of matter.

Insulator — A material that generally keeps electrons from moving through it.

Kinetic energy — Energy of motion.

Longitudinal wave — A kind of wave in which the particles go back and forth in the same direction as the wave.

Magnetic domains — Tiny magnets making up a magnetic material.

Magnetic field — A field that exerts forces on objects made of magnetic materials; made up of many lines of force.

Magnetic materials — Materials that are attracted by magnets; contain magnetic domains.

Mechanical energy — Occurs when force is applied to move or cause motion, like the wind moving a windmill.

Overtones — One of the qualities of sounds that is caused by extra vibrations.

Pitch — The "highness" or "lowness" of a tone, as on a musical scale, which is determined by the frequency of vibrations.

Potential energy — Stored energy because of its position or condition.

Prism — A triangular piece of a transparent material such as glass that separates light into its different colors.

Reflection — Bouncing off, as light bouncing off a smooth surface.

Refraction — The bending of light when it goes from one substance to another.

Sound energy — The movement of energy through a substance in the form of longitudinal waves.

Static electricity — Accumulation of positive or negative charges on an object.

Thermometer — A device for measuring temperature by means of the expansion and contraction of a liquid in a tube.

Transverse wave — A kind of wave in which the particles go back and forth crosswise to the direction of the wave.

Volts — The electrical pressure supplied to a circuit by a battery or a generator.

Wavelength — The distance from one part of a wave to the next part that is just like it.